Advances in Bioceramics and Porous Ceramics III

Advances in Bioceramics and Porous Ceramics III

*A Collection of Papers Presented at the
34th International Conference on Advanced
Ceramics and Composites
January 24–29, 2010
Daytona Beach, Florida*

Edited by
Roger Narayan
Paolo Colombo

Volume Editors
Sanjay Mathur
Tatsuki Ohji

A John Wiley & Sons, Inc., Publication

Published by John Wiley & Sons, Inc., Hoboken, New Jersey.
Published simultaneously in Canada.

For general information on our other products and services or for technical support, please contact our
Customer Care Department within the United States at (800) 762-2974, outside the United States at
(317) 572-3993 or fax (317) 572-4002.

Wiley also publishes its books in a variety of electronic formats. Some content that appears in print may
not be available in electronic format. For information about Wiley products, visit our web site at
www.wiley.com.

Library of Congress Cataloging-in-Publication Data is available.

ISBN 978-0-470-59471-1

Printed in the United States of America.

10 9 8 7 6 5 4 3 2 1

Contents

BIOCERAMICS

Preface

This issue contains the proceedings of the "Porous Ceramics: Novel Developments and Applications" and "Next Generation Bioceramics" symposia, which were held on January 24–29, 2010 at the Hilton Daytona Beach Resort and the Ocean Center in Daytona Beach, Florida, USA.

The interaction between ceramic materials and living organisms is a leading area of ceramics research. Novel bioceramic materials are being developed that will provide improvements in diagnosis and treatment of medical and dental conditions. In addition, bioinspired ceramics and biomimetic ceramics have generated considerable interest in the scientific community. The "Next Generation Bioceramics" symposium addressed several leading areas related to the development and use of novel bioceramics, including advanced processing of bioceramics; biomineralization and tissue-material interactions; bioinspired and biomimetic ceramics; ceramics for drug delivery; ceramic biosensors; in vitro and in vivo characterization of bioceramics; mechanical properties of bioceramics; and nanostructured bioceramics. This symposium promoted lively interactions among various stakeholders in bioceramics development, including scientists from academia, industry, and government.

The "Porous Ceramics" symposium aimed to bring together engineers and scientists in the area of ceramic materials containing high volume fractions of porosity, in which the porosity ranged from nano- to millimeters. These materials have attracted significant academic and industrial attention for use in environmental applications, an area where ceramics, particularly porous ones, play a key role because of their suitable properties. Therefore, a significant number of contributions, of which some are present in this volume, was devoted to the fabrication and characterization of porous ceramics for gas purification (e.g., H_2 separation and CO_2 separation) as well as to particulate filtration (e.g., diesel engine soot). We expect that scientific activity in this field will increase for the foreseeable future. This symposium will continue to play a significant role in encouraging interactions between researchers from academia and industry as well in promoting the dissemination of research in order to benefit the society at large.

A leading area of ceramics research involves the development of porous ceramics for medical, dental, and biotechnology applications. For example, porous ceramics are under development for use as bone substitutes because a porous structure may enhance tissue ingrowth. Therefore, tailoring of porosity to give specific characteristics, in terms of the amount of interconnecting cells and of the cell and cell window size is required. A joint session involving participants from bioceramics and porous ceramics symposia was therefore held in order to stimulate discussion and productive interactions between the two scientific communities.

We would like to thank Greg Geiger, Mark Mecklenborg, Marilyn Stoltz, Marcia Stout, Anita Lekhwani and the staff at The American Ceramic Society and John Wiley & Sons for making this proceedings volume possible. We also give thanks to the authors, participants, and reviewers of the proceedings volume. We hope that this issue becomes a useful resource for porous ceramics and bioceramics researchers in academia, government, and industry, which contributes to the advancement of ceramic science and technology and signifies the leadership of The American Ceramic Society in these rapidly developing areas.

ROGER NARAYAN
University of North Carolina and North Carolina State University

PAOLO COLOMBO
Università di Padova (Italy) and The Pennsylvania State University

Introduction

This CESP issue represents papers that were submitted and approved for the proceedings of the 34th International Conference on Advanced Ceramics and Composites (ICACC), held January 24–29, 2010 in Daytona Beach, Florida. ICACC is the most prominent international meeting in the area of advanced structural, functional, and nanoscopic ceramics, composites, and other emerging ceramic materials and technologies. This prestigious conference has been organized by The American Ceramic Society's (ACerS) Engineering Ceramics Division (ECD) since 1977.

The conference was organized into the following symposia and focused sessions:

Symposium 1	Mechanical Behavior and Performance of Ceramics and Composites
Symposium 2	Advanced Ceramic Coatings for Structural, Environmental, and Functional Applications
Symposium 3	7th International Symposium on Solid Oxide Fuel Cells (SOFC): Materials, Science, and Technology
Symposium 4	Armor Ceramics
Symposium 5	Next Generation Bioceramics
Symposium 6	International Symposium on Ceramics for Electric Energy Generation, Storage, and Distribution
Symposium 7	4th International Symposium on Nanostructured Materials and Nanocomposites: Development and Applications
Symposium 8	4th International Symposium on Advanced Processing and Manufacturing Technologies (APMT) for Structural and Multifunctional Materials and Systems
Symposium 9	Porous Ceramics: Novel Developments and Applications
Symposium 10	Thermal Management Materials and Technologies
Symposium 11	Advanced Sensor Technology, Developments and Applications

Focused Session 1 Geopolymers and other Inorganic Polymers
Focused Session 2 Global Mineral Resources for Strategic and Emerging
                   Technologies
Focused Session 3 Computational Design, Modeling, Simulation and
                   Characterization of Ceramics and Composites
Focused Session 4 Nanolaminated Ternary Carbides and Nitrides (MAX Phases)

The conference proceedings are published into 9 issues of the 2010 Ceramic Engineering and Science Proceedings (CESP); Volume 31, Issues 2–10, 2010 as outlined below:

- Mechanical Properties and Performance of Engineering Ceramics and Composites V, CESP Volume 31, Issue 2 (includes papers from Symposium 1)
- Advanced Ceramic Coatings and Interfaces V, Volume 31, Issue 3 (includes papers from Symposium 2)
- Advances in Solid Oxide Fuel Cells VI, CESP Volume 31, Issue 4 (includes papers from Symposium 3)
- Advances in Ceramic Armor VI, CESP Volume 31, Issue 5 (includes papers from Symposium 4)
- Advances in Bioceramics and Porous Ceramics III, CESP Volume 31, Issue 6 (includes papers from Symposia 5 and 9)
- Nanostructured Materials and Nanotechnology IV, CESP Volume 31, Issue 7 (includes papers from Symposium 7)
- Advanced Processing and Manufacturing Technologies for Structural and Multifunctional Materials IV, CESP Volume 31, Issue 8 (includes papers from Symposium 8)
- Advanced Materials for Sustainable Developments, CESP Volume 31, Issue 9 (includes papers from Symposia 6, 10, and 11)
- Strategic Materials and Computational Design, CESP Volume 31, Issue 10 (includes papers from Focused Sessions 1, 3 and 4)

The organization of the Daytona Beach meeting and the publication of these proceedings were possible thanks to the professional staff of ACerS and the tireless dedication of many ECD members. We would especially like to express our sincere thanks to the symposia organizers, session chairs, presenters and conference attendees, for their efforts and enthusiastic participation in the vibrant and cutting-edge conference.

ACerS and the ECD invite you to attend the 35th International Conference on Advanced Ceramics and Composites (http://www.ceramics.org/icacc-11) January 23–28, 2011 in Daytona Beach, Florida.

Sanjay Mathur and Tatsuki Ohji, Volume Editors
July 2010

Bioceramics

BIODEGRADABLE RARE EARTH LITHIUM ALUMINOBORATE GLASSES FOR BRACHYTHERAPY USE

J.E. White[1], D. E. Day[2], R.F. Brown[3] and G.J. Ehrhardt[4]
1. Praxair Inc., Tonawanda, NY 14150
2. Department of Materials Science and Engineering,
3. Department of Biological Sciences, Missouri University of Science and Technology, Rolla, MO 65409
4. University of Missouri Research Reactor, University of Missouri-Columbia, MO 65211

ABSTRACT

Rare earth lithium aluminoborate (RELAB) glasses containing dysprosium (Dy) or holmium (Ho) were investigated to determine their suitability for irradiating diseased organs in the body; i.e., brachytherapy. The chemical durability of a glass used for brachytherapy is especially important since none of the radioactive Dy or Ho should escape from the glass, but in some applications the glass should eventually degrade in the body. The chemical durability (dissolution rate) of Dy or Ho containing glass microspheres (20 to 40 microns in diameter) immersed in phosphate buffered saline (PBS) or DI water at $37^{\circ}C$ was measured as a function of their alumina content, 5 to 20 mole percent. A highly desirable aspect of the Dy or Ho lithium aluminoborate glasses is that they degraded in PBS or DI water in a non-uniform way such that the radioisotope was confined to the remnants of the corroded microsphere. Dy microspheres injected into the stifle (knee) joint of a rat were noticeably corroded after two weeks. The *in-vivo* results indicate that RELAB microspheres can safely deliver localized doses of beta radiation to diseased organs and should eventually be cleared from the body.

INTRODUCTION

The concept of using radioactive glass microspheres to irradiate organs in the body; i.e., brachytherapy is not new. Microspheres made from a chemically durable yttria aluminosilicate glass (YAS), containing beta emitting ^{90}Y, have been used[1] for several years to treat patients with inoperable liver cancer with good results. The YAS glass was developed to have a high chemical durability in the body, but there are other applications where a biodegradable radioactive glass microsphere could be used[2] such as in the treatment of arthritic joints; what is called radiation synovectomy.

Colloidal particles containing radioactive ^{90}Y and ^{198}Au have been used in Europe for radiation synovectomy[3] along with radiocolloids containing ^{165}Dy, ^{166}Ho, and ^{32}P. A problem with radiocolloids is their small size, typically 10 to 100nm, which enables some of the radioactive particles (5 to 25% of the initial dose) to escape from the treated joint. Biodegradable glass microspheres of a larger size, 1 to 20 μm, are expected to reduce the undesirable leakage of radiation from the joint and, depending upon their composition, also have the advantage that the as-made microspheres can be made radioactive by neutron activation. This reduces greatly the need to handle radioactive materials.

In the present study, lithium aluminoborate glasses containing either Dy_2O_3 (DyLAB) or Ho_2O_3 (HoLAB) were chosen for study since (1) ^{165}Dy and ^{166}Ho have been used previously[4,5] for radiation synovectomy, (2) both radioisotopes are neutron activatable and (3) these glasses were expected to gradually degrade in the body[2,6]. The chemical durability of these glasses and their degradability in

3

simulated body fluids was of greatest interest since it is important that no appreciable radioactivity escape from the site being treated.

EXPERIMENTAL PROCEDURE

Glass melting and microsphere preparation

Fifty grams of each DyLAB and HoLAB composition listed in Table I were prepared by melting a homogeneous mixture of high purity Dy_2O_3, Ho_2O_3, Li_2CO_3, Al_2O_3, H_3BO_3, and SiO_2 in a platinum-rhodium crucible at temperatures between 1200 and 1300 $^\circ$C for 1hr in an electric furnace. Each melt was stirred, cast into metal molds and annealed. Portions of each annealed RELAB glass were crushed to particles smaller than 45 μm which were spheroidized using a propane/air flame. The DyLAB glasses that were neutron activated to form ^{165}Dy were made using Li_2CO_3 enriched in ^{7}Li and H_3BO_3 enriched in ^{11}B.

Chemical durability measurements

Samples (annealed or unannealed) of the DyLAB and HoLAB glasses were placed in separate polyethylene vials and immersed in phosphate buffered saline (PBS)* solution, pH 7.4, at 37 $^\circ$C. The glass surface area to solution volume ratio (SA/V ratio) was ~0.1 cm^{-1}. Periodically, the samples were removed from the PBS, washed, dried and weighed. This procedure was repeated for periods up to 90 days. The dissolution rate (DR) of each glass for a given time period was calculated by dividing the measured weight loss by the sample's surface area and the elapsed time.

The amount of Dy released by microspheres (28 μm average diameter) of the DyLAB glasses immersed in the PBS solution at 37°C was determined by analyzing the PBS solution by inductively coupled plasma spectroscopy (ICP). Approximately 70 mg of DyLAB-5, -10, -15, and -20 glass microspheres were each immersed in 30 ml of PBS (SA/V ~ 1.8 cm^{-1}) for 5 and 24 hrs. In one case, the DyLAB-10 microspheres remained immersed in the PBS solution for 23 days. Each PBS solution was analyzed with a lower detection limit of 0.1 ppm for Dy, B, Al, Si and Mg. The solutions were not analyzed for Li.

In another experiment, radioactive microspheres of the DyLAB-10 and -20 glass, which had been neutron activated to form ^{165}Dy, were immersed in deionized (DI) water for up to 10 hrs at 37 $^\circ$C. The radioactivity of the microspheres and the decanted water samples was measured by counting the 94 keV gamma radiation emitted by the ^{165}Dy (half-life of 2.33 hrs). At three hour intervals, a sample of the DI water was removed (and replaced with fresh DI water), filtered to remove any radioactive glass microspheres, and its activity measured with a scintillation counter. The radioactivity of the DyLAB microspheres in the DI water was measured as a control.

The surface composition of the as-made and corroded DyLAB glasses was analyzed by Electron dispersion spectroscopy (EDS) and by x-ray photoelectron spectroscopy (XPS), but those results are not discussed herein.

Animal (in-vivo) experiments

Non radioactive DyLAB-10 microspheres, 20 to 25 μm in diameter, were injected into the stifle (knee) joint of a healthy rat to assess their in-vivo biocompatibility and degradation characteristics. The DyLAB-10 microspheres were suspended in a glycerol-saline solution and 2 and 4 mg of microspheres (roughly 50,000 spheres/mg) and injected into the left and right stifle joint, respectively. After two weeks, the animal was sacrificed, the stifle joints were excised, fixed in formalin, dehydrated, and embedded in polymethymethacrylate (PMMA). The PMMA-embedded samples were sectioned with a diamond wafering blade, ground and polished to an optical finish, and stained for histological evaluation as described elsewhere[2]. The stained sections were examined by optical microscopy to determine the condition and location of the microspheres as well as limited evaluation of tissue responses.

RESULTS

Glass formation

All of the DyLAB and HoLAB compositions in Table 1 formed bubble-free, pale yellow or orange glasses, respectively, with no noticeable tendency for devitrification. The smooth surface and size uniformity of the DyLAB-5 microspheres shown in Figure 1 is representative of the microspheres after flame spheroidization.

Chemical durability

The percent weight loss/cm^2 for plates of the HoLAB (2 mol percent Ho_2O_3) and DyLAB (5 mol percent Dy_2O_3) glasses immersed in PBS at 37 $^\circ$C is shown in Figures 2 and 3, respectively. In both cases, the percent weight loss decreased with increasing alumina content, but the DyLAB glasses are clearly more durable than comparable HoLAB glasses up to an alumina content of 20 % where the weight loss/cm^2 for both glasses is essentially the same, see Figure 3. The most likely explanation for the smaller weight loss for the DyLAB glasses is the higher percentage of Dy_2O_3 (5 mol %) compared to the lower percentage of Ho_2O_3 (2 mol %).

The average dissolution rates (DR) in PBS at 37 $^\circ$C, calculated from the weight loss data, for the DyLAB and HoLAB glasses are compared in Figure 4 and clearly show a similar dependence upon the alumina content. Note also that the DR calculated for short times, 7 days, and that calculated for much longer times, 75 days for DyLAB and 90 days for HoLAB, are essentially the same. In other words, the weight loss of each glass continues to increase essentially linearly with time up to the longest time measured.

Additional important evidence for the corrosion of the DyLAB glasses, was provided by the ICP analysis of the PBS solution (at 37 $^\circ$C) in which DyLAB glass microspheres were immersed for 24 hrs. While the amounts of B, Al, Mg, and Si in the PBS after 24 hrs ranged from a low of ~ 1 ppm for Si to a high of 18 ppm for Mg for the least durable DyLAB-5 glass, the amount of Dy was below the detection limit of 0.1 ppm Dy. Even in the case where microspheres of the DyLAB-10 glass were immersed in the PBS solution for 23 days, the amount of Dy was less than 0.1 ppm. These results indicate that the DyLAB glasses and, most probably, the HoLAB glasses degrade in a non-uniform fashion in PBS in such a way that all of the glass constituents are released, with the notable exception of the rare earth ions. The PBS solution was not analyzed for Li, but it was assumed to be present.

The results shown in Figure 5 for radioactive DyLAB-10 and -20 glass microspheres immersed in DI water (at 37 $^\circ$C) for ~ 11 hrs, during which time 96.4 % of the ^{165}Dy will have decayed, provide further evidence that no ^{165}Dy is released from either glass composition. As shown by the dashed line for the decanted water samples, the radioactivity of the DI water was below the detection limit (sub nanoCi) for the entire time. Furthermore, the decay in the measured activity of the glass microspheres in the DI water agreed closely with what that calculated from the 2.33 hr half-life for [165]Dy. The results for these two DyLAB glasses, whether immersed in DI water or PBS solution, are consistent and show that no detectable amounts of Dy were released during the corrosion process.

Animal experiments

The general appearance of DyLAB-10 microspheres recovered after two weeks in the stifle joint of a rat is shown in Figure 6. The microspheres were found incorporated in the synovial tissue, as opposed to the synovial fluid, and no microspheres were found outside the synovial membrane or in the articular cartilage on the bone. There was no noticeable decrease in the size of the microspheres, initial diameter ~ 23 µm, but the appearance many of the microspheres in Figure 6 indicates that a corroded surface layer had formed and surrounded what is believed to be a remnant glass core.

The laboratory rats resumed normal activity immediately after injection and did not show any evidence of distress during the two week test period. There was no discernable difference in the appearance of the joints that were injected with either 2 or 4 mg of DyLAB-10 microspheres. Furthermore, there was no evidence of necrosis of the joint tissue or of any mechanical damage (wear) to any part of the joint.

DISCUSSION

Suitability for Brachytherapy Applications

In the present investigation, the Dy and Ho aluminoborate glasses were of interest from the standpoint of their potential to gradually degrade in the body as opposed to the chemically durable aluminosilicate (YAS) glass microspheres which remain in the body for several months or longer. In applications such as radiation synovectomy of rheumatoid arthritic joints, it is considered undesirable for the radioactive particles/microspheres to remain indefinitely in the joint to avoid the possibility of damaging the joint. The DyLAB and HoLAB compositions investigated were chosen on the basis (a) Dy and Ho can be neutron activated to form desirable radioisotopes (b) none of the other elements become radioactive by neutron activation when enriched [7]Li and [11]B compounds are used as the source of Li and B (c) the concentration of Dy_2O_3 and Ho_2O_3 exceeded the minimum needed, 3 and 0.5 mol %, respectively, so that ~ 15 mg of glass would be sufficient to deliver the desired (~ 100 Gy) radiation dose, and (d) their anticipated biocompatibility and biodegradability.

The weight loss data in Figures 2 and 3 clearly show that the corrosion of the DyLAB and HoLAB glasses in PBS at 37 °C (body temperature) can be controlled over wide limits by changing the alumina content. Since the half-life of the [165]Dy and [166]Ho radioisotopes is 2.33 and 26.9 hrs, respectively, one is only concerned about the release of these radioisotopes from the glass for 24 hrs and 11.2 days, respectively, since 99.9% of the [165]Dy or [166]Ho radiation will have decayed during that time period. A comparison of the corrosion that occurs in 24 h for DyLAB glass microspheres immersed in PBS is shown in Figure 7. The visible surface cracks (likely occurred during drying) for the DyLAB-5 and -10 microspheres are evidence of corrosion occurring in 24 hrs, but the higher alumina-containing DyLAB-15 and -20 microspheres show no noticeable signs of corrosion in 24 hrs. Based on their higher weight loss and dissolution rate, comparable HoLAB glasses are expected to degrade more rapidly in PBS than DyLAB glasses of equivalent alumina content.

The most significant finding for the DyLAB glasses was that when they were immersed in PBS, the solution contained significant amounts of the elements in the glass, with the exception of Dy whose concentration was below the ICP detectability limit of 0.1 ppm. This was true for even for the less durable DyLAB-5 glass which was visibly corroded in 24 hrs, see Figure 7. Even when the more durable DyLAB-10 glass was immersed in the PBS solution for 23 days, the Dy concentration was still below 0.1 ppm. It is clear from these results that the DyLAB glasses are corroded in PBS in such a way that the Dy remains trapped in the remnants of the corroded microsphere.

Similar results were obtained when radioactive DyLAB-10 and -20 glass microspheres were immersed in DI water at 37 $^{\circ}$C for ~11 h, see Figure 5. If any radioactive ^{165}Dy had been released from the glass microspheres, the DI water samples would have been radioactive, but again the activity of the water was below the detectability limit of 1 nCi. These results are strong evidence that radioactive ^{165}Dy remains trapped in the corroded microsphere.

The preceding results agree with those obtained[2] for microspheres (5 to 15 microns in diameter) made from a significantly less durable alumina-free lithium borate glass, mol % composition of 1.7 Dy_2O_3 – 24.6 Li_2O – 73.7 B_2O_3. Microspheres of this glass were immersed in a simulated synovial fluid (0.3 % hyaluronic acid dissolved in PBS) at 37 $^{\circ}$C for up to 64 days. The amounts of Dy, Li, and B in solution were determined by atomic absorption spectroscopy. After 64 days, 97 % of the lithium, and 90% of the boron had dissolved from the as-made microspheres, contrasted to only 0.3% of the dysprosium. After only 3 and 6 hrs in the simulated synovial fluid (SSF), the microspheres had a weight loss of 50 and 75%, respectively. Even so, < 0.1% of radioactive ^{165}Dy was detected in the SSF when radioactive microspheres of this glass were immersed in the SSF solution for 7 hrs.

The in-vitro measurements from the present work and other studies[2,6] show that microspheres made from dysprosium-containing borate glasses, with and without alumina, undergo significant corrosion and degradation in PBS, DI water or SST at normal body temperature, but negligible amounts of dysprosium are released from the corroded remains of the microspheres.

In-vivo degradation

The in-vivo experiments in this study were limited, but the degradation observed for DyLAB-10 microspheres incorporated into the synovial tissue of a laboratory rat at two weeks, Figure 6, was consistent with the corrosion observed for this glass in PBS and DI water. The microspheres in Figure 6 show obvious signs of degradation, but it is important to note that the size of the microspheres has not decreased noticeably. When used for radiation synovectomy, biodegradable microspheres whose size does not decrease during degradation should be advantageous since they should be less likely to escape from the joint than a microsphere which became smaller during degradation.

Another example[2] of the biodegradation of lithium borate glass microspheres containing only 1.7 mol % Dy_2O_3, compared to 5 mol % Dy_2O_3 in the present work, is shown in Figure 8. After 16 weeks in the stifle joint of a rat, these microspheres (5 to 15 µm in diameter) are more extensively degraded as expected, but there has been no noticeable change in size. This is the same glass which had a 75 % weight loss after only 6 hr in simulated synovial fluid. As in the present work, the microspheres in Figure 6 were found incorporated into the synovial soft tissue and were never found in contact with the articular cartilage or bone.

It important to note that in the present work and in reference 2, where biodegradable Dy-containing borate glass microspheres were injected into the stifle joint of rats, the microspheres were found incorporated within the synovial tissue rather than suspended within the synovial fluid. In another long term study[7], five mg of nonradioactive, chemically insoluble samarium aluminosilicate glass microspheres (10 to 20 μm in diameter) were injected into the stifle joint of rabbits and these microspheres were also found incorporated in the synovial tissue of the joint. Initially, a slight foreign body response (mild inflammation) was noted, but no damage was detected in joints which contained microspheres for up to 1 year. The fact that the microspheres become incorporated in the synovial tissue fairly quickly may partially explain the absence of any mechanical damage to the articular cartilage and bone.

The animal experiments clearly demonstrate the biodegradability of DyLAB glasses. It is also apparent from the corrosion experiments conducted in PBS, DI water and SSF, that these three liquids are good predictors, at least for the glasses tested, of the degradation likely to occur in physiological fluids. Thus, there is no reason to expect that harmful amounts of radioactive ^{165}Dy would escape from DyLAB glasses as they are degraded by physiological liquids in the body.

HoLAB glasses were not studied in the same detail as the DyLAB glasses in the present work, but because of their similar reaction in PBS the results for the DyLAB glasses are assumed to be a reliable indicator of the behavior of HoLAB glasses and other rare earth containing (e.g., Y, Sm, Nd, Pr, Tb) lithium aluminoborate glasses of similar composition. The amount of ^{166}Ho and other rare earth radioisotopes released from such glasses as they degrade in physiological fluids is expected to be negligible.

Corrosion of RELAB glasses

The DyLAB and HoLAB glasses immersed in PBS or DI water in the present study obviously degrade in a non-uniform fashion such that the Dy and Ho radioisotopes remain trapped within the remnants of the corroded glass. Real time video microscopy[6] of DyLAB microspheres immersed in PBS shows that a reaction starts at the external surface, see Figure 9, and progresses inwardly at a velocity dependent upon the glass composition and the liquid composition, pH and temperature. An example of the progression of this reaction for DyLAB-10 microspheres immersed in PBS is shown by a shrinking core in Figure 9 which is visible at 49 hrs, becomes progressively smaller at 99 hrs and is barely visible in many of the microspheres at 160 hrs. The reaction rate calculated from the diameter of the shrinking core for the DyLAB-10 and the more reactive DyLAB-5 glass was 0.07 and 0.30 μm per hr, respectively. Note that the diameter of the corroding microspheres in Figure 9 does not change during this reaction.

A model proposed[8] for the degradation or non-uniform corrosion of DyLAB and HoLAB glasses in PBS or DI water is shown in Figure 10. Initially, a large fraction of the soluble components such as B, Li, and Mg, along with smaller amounts of the less soluble components such as Al and Si, are leached from the glass surface and go into solution forming a corroded surface layer which is enriched in the less soluble components such as the rare earth, Al and Si. In water, the less soluble components are likely present in an amorphous hydrated layer, but in liquids such as PBS which contain phosphate, the rare earth cations have been found to react with the phosphate anions and form insoluble rare earth phosphates[2,6,8]. The formation of insoluble rare earth hydroxides or phosphate compounds prevents the rare earth cations from escaping from the degrading glass. This process continues until the microsphere is fully reacted.

The two important aspects of this degradation process, from the standpoint of brachytherapy applications, is that (1) the size and shape of the glass object (microsphere/fiber/other) does not change during degradation, and (2) cations, such as the rare earth radioisotopes, which react to form insoluble products remain trapped in the corroded material, at least for the time they are radioactive. These two features of the degradation process are particularly useful for brachytherapy applications where it is important that none of the radioisotopes escape from the glass particle.

CONCLUSIONS

Rare earth lithium aluminoborate glasses are considered good candidates for brachytherapy applications. They can contain large amounts of neutron activatable radioisotopes of varying half-life and radioemission characteristics (beta vs gamma, energy, range in tissue, etc) so that therapeutic doses of localized radiation can be delivered and tailored to a particular organ or treatment site. The results for the DyLAB and HoLAB glasses in this work indicate that they are degradable in physiological liquids and the rare earth radioisotopes remain trapped in the degraded remnants. Since the rare earth elements have similar properties, it is expected that most rare earth containing borate or aluminoborate glasses will react with physiological liquids in a manner similar to that of the Dy glasses and the rare earth radioisotope will be trapped in the corroded remnants.

Of the glasses studied in the present work, the DyLAB-5 and DyLAB-10 glasses are considered the best biodegradable candidates at this time for brachytherapy where ^{165}Dy would be useful such as radiation synovectomy of arthritic joints. In 24 hrs, essentially all (>99.9%) of the ^{165}Dy has decayed during this period and the amount of ^{165}Dy released is negligible. Furthermore, the significant biodegradation of these microspheres in 2 to 16 weeks in laboratory rats suggests that they should be completely removed from the body with essentially no radioactive ^{165}Dy escaping from the target site. As a final point, biodegradable glass microspheres 10 to 20 μm in diameter, as used in this study, are considered particularly useful for radiation synovectomy of arthritic joints since there was no evidence of any mechanical damage to the joint after even one year in rabbits. This is likely due to the fact that such microspheres become rapidly immobilized within the synovial tissue and are unavailable to irritate cartilage and other portions of the joint.

REFERENCES

1. R. Salem and K.G. Thurston, "Radioembolization with yttrium-90 Microspheres: A State-of-the-Art Brachytherapy Treatment for Primary and Secondary Liver Malignancies, Part 3: Comprehensive Literature Review and Future Direction," J. Vas. Inter. Radiol. **17** 1571-1594 (2006).

2. S. Conzone, R. Brown, D.E. Day, G. Ehrhardt, "In-Vitro and In-Vivo Dissolution Behavior of a Dysprosium Lithium Borate Glass Designed for the Radiation Synovectomy Treatment of Rheumatoid Arthritis," J. Biomedical Materials Research **60** (2) 260-68 (2002).

3. J. C. Harbert and H. A. Ziessman, "Therapy with Intraarterial Microspheres." Nuclear Medicine Annual 1987, edited by L. Freeman and H. Weissmann, 295-319.Raven Press, New York.

4. C.B. Sledge, J. Noble, D.J. Hnatowich, R. Kramer, S. Shortkroff, "Experimental Radiation Synovectomy by ^{165}Dy Ferric Hydroxide Macroaggregate," Arth. Rheum., **20** (7) 1334-42 (1977).

5. M. Neves, F. Waerenborgh and L. Patricio, "Palladium-109 and Holmium-166 Potential Radionuclides for Synoviotherapy - Radiation Absorbed Dose." Appl. Rad. Isot., **38** [9] (1987).

6. D.E. Day, J.E. White, R.F. Brown & K.D. McMenamin, "Transformation of Borate Glasses into Biologically Useful Materials," Glass Technology, **44** (2) 75-81 (2003).

7. G.J. Ehrhardt and J.C. Lattimer, University of Missouri-Columbia, personal communication.

8. D.E. Day and J.E.White, U.S. patent 6,379,648; 30 April 2002, "Biodegradable Glass Compositions & Methods for Radiation Therapy."

*Phosphate buffered saline, P-4417, Sigma Chemical Co., St. Louis, MO.

Table I. Nominal Composition (mol %) of the DyLAB and HoLAB Glasses

Glass*	RE$_2$O$_3$**	Li$_2$O	Al$_2$O$_3$	B$_2$O$_3$
DyLAB-5	5 (23.7)	16	5	66
DyLAB-10	5 (23.1)	15	10	62
DyLAB-15	5 (22.6)	14	15	58
DyLAB-20	5 (22.0)	13	20	54
HoLAB-5	2 (10.9)	17	5	68
HoLAB-10	2 (10.6)	16	10	64
HoLAB-15	2 (10.3)	15	15	60
HoLAB-20	2 (10.0)	14	20	56

*The number in each glass ID corresponds to the mole % Al$_2$O$_3$ in the glass. Each glass also contains 5 SiO$_2$ and 3 MgO, mol%.
**Dy$_2$O$_3$ (DyLAB) or Ho$_2$O$_3$ (HoLAB). Weight % in parenthesis.

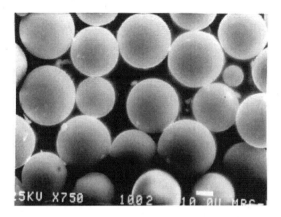

Figure 1. SEM micrograph of DyLAB-5 glass microspheres made by flame spheroidization. White scale bar is 10 μm.

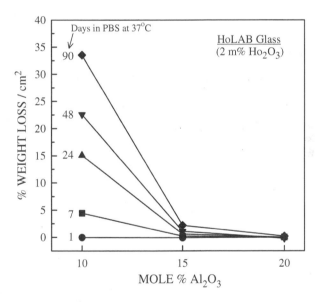

Figure 2. Percent weight loss/cm^2 (estimated) as a function of mole % Al$_2$O$_3$ for annealed HoLAB glass plates after immersion in PBS solution at 37 °C for the indicated times up to 90 days. The lines are included to guide the eye.

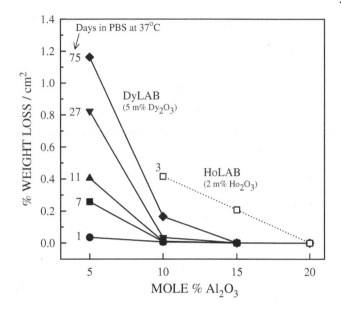

Figure 3. Percent weight loss/cm^2 as a function of mole % Al$_2$O$_3$ for annealed DyLAB glass plates after immersion in PBS solution at 37 °C for the indicated times, up to 75 days. The weight loss for HoLAB glass plates after immersion in PBS for 3 days (open squares; dashed line) are included for comparison. The lines are included to guide the eye.

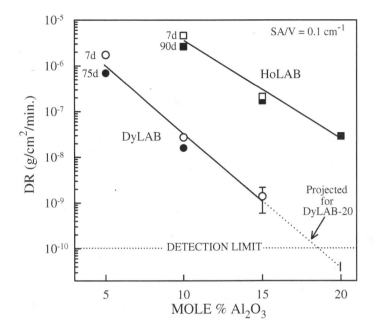

Figure 4. Calculated dissolution rate (DR) as a function of mole % Al₂O₃ for DyLAB and HoLAB glasses immersed for 7 [open symbol] or 75 days (90 days for HoLAB) [closed symbol] in PBS solution (pH 7.4) at 37 °C. Lines are included to guide the eye. The calculated experimental error in DR was approximately $\pm 1 \times 10^{-9}$ g/cm²/min.

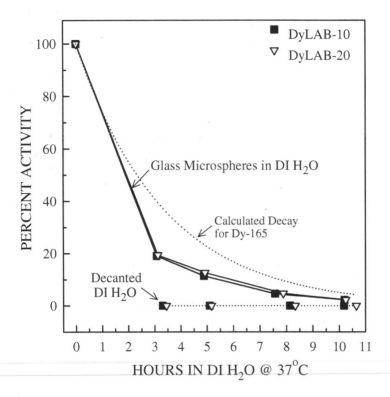

Figure 5. Percent activity as a function of time for radioactive DyLAB-10 and -20 glass microspheres (28 μm average diameter) and decanted DI water samples in which the radioactive microspheres were immersed at 37 °C for up to 11 hours. The solid lines show the decay in the measured activity of the DyLAB glass microspheres and the dashed lines show the activity calculated for the decay of [165]Dy (2.33 hr half-life) and the measured activity of the decanted DI water samples.

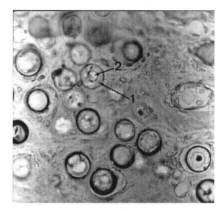

Figure 6. Photomicrograph (480x) of DyLAB-10 glass microspheres (20 to 25 μm in diameter) two weeks after injection into the healthy stifle joint of a rat. The microspheres are incorporated into the synovial soft tissue and most had started to noticeably degrade as indicated by the glass microsphere where arrow (1) points to a corroded layer which surrounds an unreacted glass core, arrow (2).

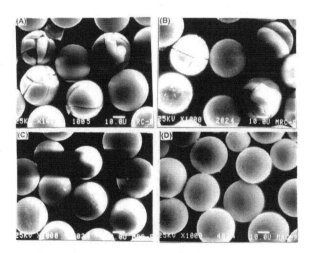

Figure 7. Representative SEM photomicrographs of (A) DyLAB-5 (B) DyLAB-10 (C) DyLAB-15 and (D) DyLAB-20 glass microspheres after immersion in PBS solution (pH 7.4) at 37 °C for 24 hours. Cracks on the surface of the DyLAB-5 and -10 microspheres, which appeared after the spheres were removed from solution, are evidence of the rapid corrosion of these two glasses. White bar is 10 μm.

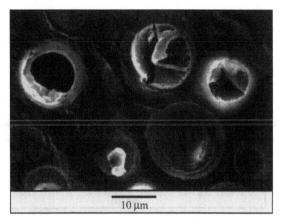

Figure 8. SEM micrograph of partially degraded dysprosium (1.7 mol% Dy_2O_3) lithium borate microspheres after 16 weeks in the stifle joint of a rat (reference 2).

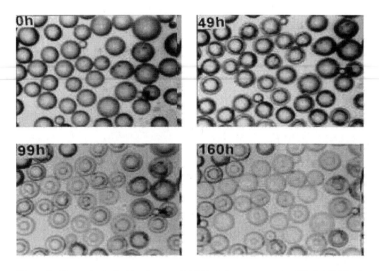

Figure 9. Chronological video images (220X) of the same group of DyLAB-10 glass microspheres (ave. diameter of 28 μm) after immersion in PBS solution (pH 7.4) at 22 °C for the time (hrs) shown in the upper left corner. A thin corroded layer at the outer surface of the microspheres is visible in the 49 hr image, becomes progressively thicker at 99 hr and at 160 hr many of the microspheres are fully corroded. Note that the size of the microspheres did not change. (Ref 6)

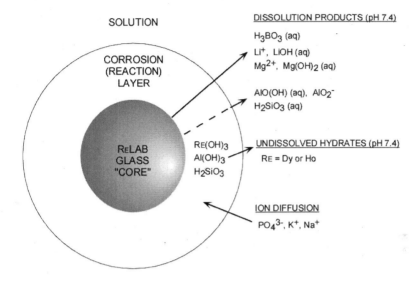

Figure 10. Schematic of a model for the non-uniform corrosion (reaction) of RELAB glass microspheres in aqueous liquids (e.g., PBS) whereby a layer rich in the rare earth and other poorly soluble cations, such as Al and Si, forms on the surface of the glass microsphere. The core of unreacted glass continues to decrease as the preferential reaction continues until the entire microsphere is reacted as a result of non-uniform (preferential) dissolution. Possible reaction products of the remaining RELAB glass components are shown along with the diffusion of ions in the PBS into the corroded layer.

ANALYTICAL MODEL FOR PREDICTION OF RESIDUAL STRESS IN ZIRCONIA-PORCELAIN BI-LAYER

M. Allahkarami, H. A. Bale, J. C. Hanan[*]

Mechanical and Aerospace Engineering, Oklahoma State University
Tulsa, OK, USA

ABSTRACT

In the case of ceramics, bi-layer residual stresses may be sufficiently large to influence the failure of a part; so, they should not be neglected in design analysis. In the case of a dental crown it is possible to analytically model the expected residual stress between the core and veneer layer, and provide a range to validate experimental results. The aim of this analytical model is to simulate the layers of a crown with clinically relevant veneer thickness including the residual stress between the core and veneer layer. For a bi-layer of 0.5 mm thickness, yttria stabilized tetragonal zirconia veneered by 0.5 mm thick porcelain gives a predicted residual stress (σ_x) of -60 MPa, 100 MPa, -40 MPa, and 10 MPa for the bottom of the zirconia layer, the zirconia at the interface, the porcelain at the interface, and the top surface of the porcelain respectively.

INTRODUCTION

It is possible to experimentally measure residual stress by different techniques, for example the hole drilling method and other cutting or slitting methods, indentation on several scales, optical or ultra-sonic methods, and X-ray diffraction. In the case of a dental crown it is possible to perform FEA on the complex geometry or idealized geometry[1] and analytically model on a simplified geometry to predict the residual stress due to the coefficient of thermal expansion mismatch between the core and veneer layer.

Owing to their high compressive strength, high toughness (for ceramics), crack resistance, and aesthetic worth; zirconia based ceramics have been employed by dentists as restoration materials. Although skillfully crafted ceramics mimic the appearance of natural teeth and should have desired mechanical properties, there are reports of failure during service [2,3,4,5]. In ceramics, thermal stresses are not avoidable. Fabrication processes require temperature changes from sintering to room temperature. The thermal stresses are not induced because of layer expansions by temperature change, but because they are not free to expand or contract. In some cases such residual stresses may be sufficiently large to influence the failure of a crown and thus should not be neglected in design analysis. Quantitative measurement of residual stresses in ceramics at the micro-scale is difficult. Polychromatic X-ray micro-diffraction provides grain orientation and residual stresses in ceramics[6,7]. Interpretation of such results has opened up a broad range of issues related to the residual stresses in these ceramics. This

---
[*] Corresponding author; e-mail: jay.hanan@okstate.edu; phone: 918-594-8238.

motivates building an analytical model to predict the residual stress and provide a background for interpretation of measured results. Timoshenko[8] derived a general solution for bi-layer bending due to residual stresses. This was recently modified for ceramics with dental applications[9,10,11].

METHOD

This analytical model used equations that couple the strain at any point of the ceramic or porcelain to the curvature and displacement of a geometric mid-plane of the two coupled plates. Figure 1 illustrates a cross section of bi-material plates before the residual stresses were in effect and after the resulting deformation due to bending[12].

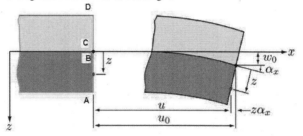

Figure 1. Cross section of bi material plates before bending and after deformation due to bending.[12]

It was assumed initially that line ABCD is straight and perpendicular to the mid plane and remains straight and perpendicular to mid plane after deformation. Displacements of point B after deformation in directions x, y, and z will be u_0, v_0, and w_0 respectively. The slope of the bi-material mid plane in the x direction is:

$$\alpha = \frac{\partial w_0}{\partial x} \tag{1}$$

Where u is displacement in the x direction of a point C located on line ABCD at distance z from the mid-plane which is given by,

$$u = u_0 - z\alpha \tag{2}$$

Combining equation (1) and (2) we obtain,

$$u = u_0 - z\frac{\partial w_0}{\partial x}; \quad v = v_0 - z\frac{\partial w_0}{\partial y} \tag{3}$$

Equation (3) relates displacement u in the x direction of an arbitrary point to the distance z from the mid-plane. Similarly for the y direction it relates displacement v to the distance z from the geometrical mid plane. Using the definition of strain and neglecting normal strain, ε_z,

$$\begin{Bmatrix} \varepsilon_x \\ \varepsilon_y \\ \gamma_{xy} \end{Bmatrix} = \begin{Bmatrix} \varepsilon_x^0 \\ \varepsilon_y^0 \\ \gamma_{xy}^0 \end{Bmatrix} + z \begin{Bmatrix} k_x \\ k_y \\ k_{xy} \end{Bmatrix} \tag{4}$$

Where, ε_x^0, ε_y^0 and γ_{xy}^0 are the first derivatives of the displacements u and v; k_x, k_y, and k_{xy} are the second derivatives of u and v. Equation (4) shows a linear relation between strains and thickness variation in bi-materials. For two dimensional cases, all the terms related to the z-axis may be dropped to simplify the stress-strain relation to,

$$\begin{Bmatrix} \sigma_x \\ \sigma_y \\ \tau_{xy} \end{Bmatrix} = \begin{bmatrix} \bar{Q}_{11} & \bar{Q}_{12} & \bar{Q}_{16} \\ \bar{Q}_{12} & \bar{Q}_{22} & \bar{Q}_{26} \\ \bar{Q}_{16} & \bar{Q}_{26} & \bar{Q}_{66} \end{bmatrix} \begin{Bmatrix} \epsilon_x \\ \varepsilon_y \\ \gamma_{xy} \end{Bmatrix} \tag{5}$$

Stiffness matrixes for zirconia and porcelain layers are[2],

$$\begin{bmatrix} Q_{11} & Q_{12} & 0 \\ Q_{12} & Q_{22} & 0 \\ 0 & 0 & Q_{66} \end{bmatrix}_{zirconia} = \begin{bmatrix} 215.43 & 47.34 & 0 \\ 47.34 & 215.43 & 0 \\ 0 & 0 & 84.02 \end{bmatrix} (GPa)$$

$$\begin{bmatrix} Q_{11} & Q_{12} & 0 \\ Q_{12} & Q_{22} & 0 \\ 0 & 0 & Q_{66} \end{bmatrix}_{porcelain} = \begin{bmatrix} 73.56 & 16.18 & 0 \\ 16.18 & 73.56 & 0 \\ 0 & 0 & 23.69 \end{bmatrix} (GPa)$$

Mechanical strains denoted by ε^M while $\varepsilon_x^T, \varepsilon_y^T$ and γ_{xy}^T are thermal strain.

$$\begin{Bmatrix} \varepsilon_x^M \\ \varepsilon_y^M \\ \gamma_{xy}^M \end{Bmatrix} = \begin{Bmatrix} \varepsilon_x \\ \varepsilon_y \\ \gamma_{xy} \end{Bmatrix} - \begin{Bmatrix} \varepsilon_x^T \\ \varepsilon_y^T \\ \gamma_{xy}^T \end{Bmatrix} = \begin{Bmatrix} \varepsilon_x^0 \\ \varepsilon_y^0 \\ \gamma_{xy}^0 \end{Bmatrix} + z \begin{Bmatrix} k_x \\ k_y \\ k_{xy} \end{Bmatrix} - \begin{Bmatrix} \alpha_x \Delta T \\ \alpha_y \Delta T \\ \alpha_{xy} \Delta T \end{Bmatrix} \tag{6}$$

The coefficients of thermal expansion, are denoted by α_x and α_y.

Equation (5) is the reduced form of the following general matrix, where A, B and D, are the stiffness matrices, coupling stiffness matrix, and bending stiffness matrix, respectively.

$$\begin{Bmatrix} N \\ M \end{Bmatrix} = \begin{bmatrix} A & B \\ B & D \end{bmatrix} \begin{Bmatrix} \epsilon^0 \\ k \end{Bmatrix} \tag{7}$$

Residual stresses originate within the composite upon firing the porcelain layer on the zirconia at a temperature of 1200 K and cooling to room temperature (298 K). The calculation may be carried out in the following sequence: $\Delta T = 298 - 1200 = -902$

$$\left\{\begin{matrix} N_x^T \\ N_y^T \\ N_{xy}^T \end{matrix}\right\} = \left\{\begin{matrix} -1.7524 \\ -1.7524 \\ 0 \end{matrix}\right\} \text{ (GPa·mm)}; \left\{\begin{matrix} M_x^T \\ M_y^T \\ M_{xy}^T \end{matrix}\right\} = \left\{\begin{matrix} -0.2301 \\ -0.2301 \\ 0 \end{matrix}\right\} \text{ (GPa·mm}^2)$$

Using the inverted form of Equation (7),

$$\left\{\begin{matrix} \epsilon^0 \\ k \end{matrix}\right\} = \begin{bmatrix} A' & B' \\ B' & D' \end{bmatrix} \left\{\begin{matrix} N^T \\ M^T \end{matrix}\right\} \tag{8}$$

Using Equation (8), the matrix mid-plane strains and plate curvatures were evaluated.

$$\left\{\begin{matrix} \varepsilon_x^0 \\ \varepsilon_{xy}^0 \\ \gamma_{xy}^0 \end{matrix}\right\} = \left\{\begin{matrix} -9.8 \\ -9.8 \\ 0 \end{matrix}\right\} \times 10^{-3} \text{ and } \left\{\begin{matrix} k_x \\ k_y \\ k_{xy} \end{matrix}\right\} = \left\{\begin{matrix} -1.3 \\ -1.3 \\ 0 \end{matrix}\right\} \times 10^{-3}$$

Mechanical strains that cause the residual stresses are calculated in accordance with equation (6)

$$\left\{\begin{matrix} \varepsilon_x^M \\ \varepsilon_{xy}^M \\ \gamma_{xy}^M \end{matrix}\right\}_{zirconia} = \left\{\begin{matrix} 0.37 - 1.3z \\ 0.37 - 1.3z \\ 0 \end{matrix}\right\} \times 10^{-3} \text{ and } \left\{\begin{matrix} \varepsilon_x^M \\ \varepsilon_{xy}^M \\ \gamma_{xy}^M \end{matrix}\right\}_{porcelain} = \left\{\begin{matrix} -0.53 - 1.3z \\ -0.53 - 1.3z \\ 0 \end{matrix}\right\} \times 10^{-3}$$

The residual stress distribution was obtained by substituting the strains in the equation (5),

$$\left\{\begin{matrix} \sigma_x^T \\ \sigma_y^T \\ \tau_{xy}^T \end{matrix}\right\} = \begin{bmatrix} \bar{Q}_{11} & \bar{Q}_{12} & \bar{Q}_{16} \\ \bar{Q}_{12} & \bar{Q}_{22} & \bar{Q}_{26} \\ \bar{Q}_{16} & \bar{Q}_{26} & \bar{Q}_{66} \end{bmatrix} \left\{\begin{matrix} \varepsilon_x^0 z + k_x - \alpha_x \Delta T \\ \varepsilon_y^0 z + k_y - \alpha_y \Delta T \\ \gamma_{xy}^0 z + k_{xy} - \alpha_{xy} \Delta T \end{matrix}\right\} \tag{9}$$

The stress profile is assumed linear across the thickness of a layer. Stresses at a surface layer and interface between two layers were calculated as follows,

Zirconia layer $z = 0.5$ mm

$$\left\{\begin{matrix} \varepsilon_x^M \\ \varepsilon_{xy}^M \\ \gamma_{xy}^M \end{matrix}\right\}_{z=+0.5} = \left\{\begin{matrix} -2.8 \\ -2.8 \\ 0 \end{matrix}\right\} \times 10^{-4} \text{ and } \left\{\begin{matrix} \sigma_x^T \\ \sigma_y^T \\ \tau_{xy}^T \end{matrix}\right\} = \left\{\begin{matrix} -73.5 \\ -73.5 \\ 0 \end{matrix}\right\} \text{ (MPa)}$$

Zirconia layer $z = 0$ mm

$$\left\{\begin{matrix} \varepsilon_x^M \\ \varepsilon_{xy}^M \\ \gamma_{xy}^M \end{matrix}\right\}_{z=0} = \left\{\begin{matrix} 3.7 \\ 3.7 \\ 0 \end{matrix}\right\} \times 10^{-4} \text{ and } \left\{\begin{matrix} \sigma_x^T \\ \sigma_y^T \\ \tau_{xy}^T \end{matrix}\right\} = \left\{\begin{matrix} 97.2 \\ 97.2 \\ 0 \end{matrix}\right\} \text{ (MPa)}$$

Porcelain layer $z = 0$ mm

$$\begin{Bmatrix} \varepsilon_x^M \\ \varepsilon_{xy}^M \\ \gamma_{xy}^M \end{Bmatrix}_{\text{porcelain } z=0} = \begin{Bmatrix} 5.3 \\ 5.3 \\ 0 \end{Bmatrix} \times 10^{-4} \text{ and } \begin{Bmatrix} \sigma_x^T \\ \sigma_y^T \\ \tau_{xy}^T \end{Bmatrix} = \begin{Bmatrix} -47.6 \\ -47.6 \\ 0 \end{Bmatrix} (\text{MPa})$$

Porcelain layer $z = -0.5$ mm

$$\begin{Bmatrix} \varepsilon_x^M \\ \varepsilon_{xy}^M \\ \gamma_{xy}^M \end{Bmatrix}_{\text{porcelain } z=-0.5} = \begin{Bmatrix} 11.8 \\ 11.8 \\ 0 \end{Bmatrix} \times 10^{-4} \text{ and } \begin{Bmatrix} \sigma_x^T \\ \sigma_y^T \\ \tau_{xy}^T \end{Bmatrix} = \begin{Bmatrix} 9.7 \\ 9.7 \\ 0 \end{Bmatrix} (\text{MPa})$$

RESULTS

For x-y reference axes, the variation of residual stresses across the ceramic layer thicknesses is shown in Figure 2. Residual stress has a self equilibrating nature, so the net area in each plot and the moment of the area about any point are zero. The residual stresses were tensile on the zirconia side of the interface and compressive on the porcelain side.

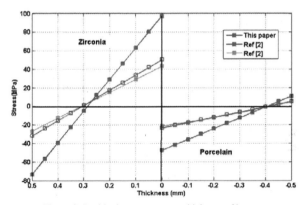

Figure 2. Residual stresses versus thickness of layers.

The slope of the residual stress curves depends on the temperature gradient from the firing temperature to the room temperature. Figure 3 shows the predicted residual stress profile for different curing temperatures. Using this model, it is possible to predict the effect of porcelain layer thickness (for fixed zirconia layer at 0.5 mm) on residual stress in particular positions of interest as shown in Figure 4 for bottom of the zirconia layer, interface of the zirconia side, interface of the porcelain side, and top surface of the porcelain. An important observation to note, is the significant variation in residual stress from a tensile stress of 80 MPa to a compressive stress of -70 MPa as the thickness of porcelain layer was increased to 0.5 mm.

Furthermore, a gradual increase in residual stress was observed from -70 MPa to -20 MPa as the thickness was further increased to 4mm. Figure 4 suggests that most of the drastic changes in residual stress occur within 0.5 mm of the porcelain thickness. The case of a constant zirconia thickness with varying porcelain thickness is of greater importance due to its relevance to the dental restoration geometries.

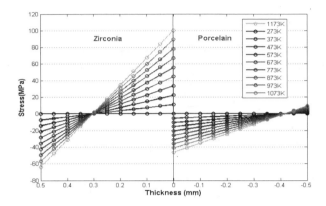

Figure 3. Predicted residual stress profile for different curing temperatures.

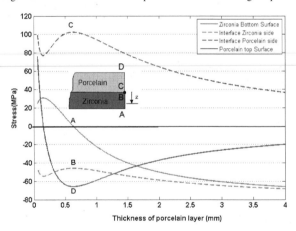

Figure 4. Residual stresses in the interface and top and bottom surface as a function of porcelain thickness. Inset image shows the locations A, B, C, and D with respect to the bilayer geometry. 'z' indicates the thickness of zirconia and porcelain.

A general result can be observed, if the thickness of both zirconia and porcelain are allowed to vary. Figure 5 illustrates predicted residual stresses as function of thicknesses in the interface, top and bottom surfaces. These simulations of the residual stresses illustrate the results across a range of clinically relevant thicknesses. This suggests there are thickness combinations where the residual stresses are maximum for the interface, particularly in the case of an interface with 0.5 mm zirconia and 0.5 mm porcelain. Independent of thickness, the porcelain top surface is always in an unfavorable state of tension.

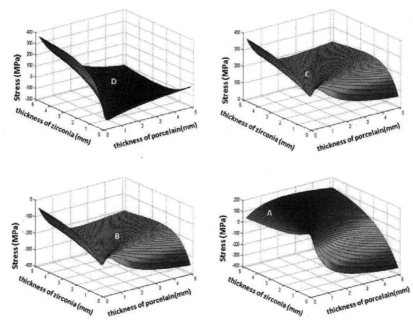

Figure 5. Variation of residual stress in interface and top and bottom surface as a function of both porcelain thickness and zirconia thickness. Locations: A, B, C, D is as indicated above in Figure 4.

CONCLUSIONS

The residual stresses between two disks of zirconia and porcelain fired together at high temperature and cooled down to room temperature were predicted. The residual stress was tensile in the zirconia side of the interface and compressive in the porcelain side. The magnitude of the residual stress depends on initial and final temperature as well as the thickness of the porcelain and zirconia layers.

Based on the fundamental equations for strain, an analytical model for a bi-layer ceramic composite was developed. The bilayer model consisting of a 0.5 mm yttria stabilized tetragonal zirconia disk veneered with a 0.5 mm thick porcelain had a predicted residual stress (σ_x) of -60 MPa, 100 MPa, -40 MPa, and 10 MPa for the bottom of the zirconia layer, the zirconia at the interface, the porcelain at the interface, and the top surface of the porcelain respectively. Using a set of generalized equations for thickness, the model simulated results of residual stress with change in thickness. Predictions of the change in stresses with thickness are non-trivial and would prove valuable in validating experimentally determined values of residual stress. Validated results could lead to recommendations for reduced stress concentration and longer service life in dental restorations.

ACKNOWLEDGEMENTS

Research support provided by the Oklahoma Health Research award (project number HR07-134), from the Oklahoma Center for the Advancement of Science and Technology (OCAST).

REFERENCES

[1] H. A. Bale, Paulo G. Coelho, N. R. Silva, V. P. Thompson, J. C. Hanan, All-Ceramic Crown Residual Stress Simulations, Material Inhomogeneity/Fine Structure Effects, Submitted to 34th International Conference on Advanced Ceramics and Composites (ICACC), (2010).

[2] M. V. Swain, Unstable cracking (chipping) of veneering porcelain on all-ceramic dental crowns and fixed partial dentures, *Acta Biomaterialia.*, **5** ,1668–1677 (2009).

[3] E. P. Butler, Transformation-toughened zirconia ceramics: critical assessment, *Material Science and Technology.*, 417 (1985).

[4] A. J. Raigrodski, Contemporary all-ceramic fixed partial dentures: a review, *Dent Clin North Am* **48**, viii (2004).

[5] A. J. Raigrodski, Contemporary materials and technologies for all-ceramic fixed partial dentures: A review of the literature, *Journal of Prosthetic Dentistry.*, **92**, 557 (2004).

[6] H. A. Bale, J. C. Hanan, Residual stresses in dental zirconia using Laue micro diffraction, *Material science, Forum* (2009).

[7] H. A. Bale, Measurement and analysis of residual stresses in zirconia dental composites using micro X-ray diffraction, Ph.D. Thesis, Oklahoma State University, Stillwater, OK, (2010).

[8] C. H. Hsueh, S. Lee, T. J. Chuang, *Journal of Applied Mechanics.*, **70** 151-154 (2003).

[9] S. Timoshenko, Analysis of Bi-Metal Thermostats, *J. Opt. Soc. Amm.*, **11**, 233–255 (1925).

[10] C. H. Hsueh, G. A. Thompson, O. M. Jadaan, A. A. Wereszczak, P. F. Becher, Analyses of layer-thickness effects in bilayered dental ceramics subjected to thermal stresses and ring-on-ring tests, *Dental materials.*, **24,** 9–17 (2008).

[11] C. H. Hsueh, C. R. Luttrell, P. F. Becher, Analyses of multilayered dental ceramics subjected to biaxial flexure tests, *Dental Materials.*, **22**, 460–469 (2006).

[12] B. D. Agarwal, L. J. Brout-man, K. Chandrashekhara, *Analysis and Performance of Fiber Composites*, Wiley, 3rd edition, 212-273 (2006).

CALCIUM-ALUMINATE BASED DENTAL LUTING CEMENT WITH IMPROVED SEALING PROPERTIES – AN OVERVIEW

Leif Hermansson, Adam Faris, Gunilla Gómez-Ortega, Emil Abrahamsson and
Jesper Lööf
Doxa Dental AB
Axel Johanssons gata 4-6, SE 75451 Uppsala, Sweden

ABSTRACT

Different types of materials are used as dental luting cements from traditional Zn-phosphate based materials and glass ionomers to resin based polymer composites. Long-term success after cementation of indirect restorations depends on retention as well as maintenance of the integrity of the marginal seal. This paper discusses a new Ca-aluminate based bioceramic as a sealing material for dental applications. The focus will be on the use as a dental luting agent for the fixation of crowns and bridges. Different aspects of the material will be presented dealing with the following main topics: biocompatibility, sealing and integration mechanisms, leakage, compressive strength, and shear bond strength between the sealing material and the surrounding tissue and crown and bridge materials. The seal is built up by an integration mechanism including a precipitation of nano-sized hydrates. The dental luting cement exhibits no shrinkage. No or minimal leakage was observed in leakage studies, which also included bacterial leakage evaluation. The bioceramic system discussed demonstrated retentive values usually associated with resin based cements and the retention was much higher than for conventional cements. Comprehensive biocompatibility studies show the material even to be *in vitro* and *in vivo* bioactive. Results from a recent one-year clinical study will also be included in this presentation. The overall result of the evaluation of this new dental luting cement is promising.

INTRODUCTION

Most dental materials employed function optimally when used at room temperature and at normal humidity, but can exhibit problems when used in the oral cavity or when in contact with teeth. According to the literature [1-5] the most common complication for all types of restoration is a collapse of the biological milieu. Bacteria, that give rise to caries, enter between the artificial material and the tooth and cause damage to it. Secondary caries is the most common cause for the need to redo both fillings and dental prosthetics.

A tooth attacked by caries can be repaired in a variety of ways. The most ideal way would be to reestablish its integrity by remineralizing, i.e. using biomimetic methods. The materials used to repair a tooth should be as close to nature´s own as possible, so that a harmony is created in the oral cavity having physical and mechanical characteristics close to those of the tooth itself. However, the acid resistance should be improved compared to that of apatite (enamel and dentine) in order to avoid caries in the dental material. The material, specifically as a dental luting cement, should also function towards the implant surface.

For materials, such as dental cement materials and other implants, that are to interact with the human body, the materials should be as bioactive or biocompatible as possible. Other properties that are required for dental cement materials are a good handling ability with simple applicability, moulding that permits good shaping ability, hardening/solidification that is sufficiently rapid for use within minutes without detrimental heat generation and that provides serviceability directly following therapy. Further required properties deal with corrosion resistance, good bonding between the cement material and biological wall and/or implant material, radio-opacity and longevity, and good aesthetics especially regarding dental filling materials.

Different types of materials are used as dental luting cements - from traditional Zn-phosphate based materials and glass ionomers to resin based polymer composites. The work presented in this paper will show how far the development has reached towards the goal mentioned above. The work will be related to the specific application – a dental bioceramic luting agent – which recently reached the European market as Ceramir^{TM} C&B.

MATERIALS AND METHODS

The material relates to a dental luting cement intended for permanent cementation of porcelain fused to metal crowns, all-metal crowns, inlays and onlays, fiber reinforced resin composite restorations and all-ceramic restorations made of high strength alumina or zirconia, as well as a cementation of an implant material to tooth structure.

The product is a hybrid between calcium aluminate and a glass ionomer cement (GIC). The glass ionomer part is essentially responsible for early properties, i.e. viscosity, setting time and early strength. The calcium aluminate contributes to basic pH during curing, minimum micro-leakage, excellent biocompatibility and long-term stability and strength. The main ingredients of the powder are: calcium aluminate, polyacrylic acid, tartaric acid, strontium-fluoro-alumino-glass and strontium fluoride.

The calcium aluminate used for the material discussed in this paper was synthesised using high purity Al_2O_3 and CaO. The correct amount of the raw materials are weighed in to a suitable container (1:1 molar ratio). The powders are intimately mixed by tumbling in excess isopropanol. Isopropanol is removed by evaporation of the solvent using an evaporator combining vacuum and heat, and finally heat-treated in an oven. The next step is filling high purity Al_2O_3 crucibles with the powder mix and heat treating it above 1350°C for the appropriate amount of time in order to get nearly mono phase calcium aluminate. After heat treatment the material is crushed using a high energy crusher. After crushing the calcium aluminate is milled to the specified particle size distribution with a $d(99)_V$ of <12µm.

Setting time, film thickness and strength are the basic important properties of a dental luting cement. The film thickness is one of the most important properties of the cement determining the size and homogeneity of the contact zone and is influenced by the glass ionomer cement (GIC) system as well as the CA system. The poly acrylic acid (PAA) has a dual function in this hybrid material. Besides functioning as in a conventional GIC being cross linked by leaking Ca^{2+} ions from both the soluble glass and the Ca-aluminate and thus building up the solid body, the poly acrylic acid also has an important role as a dispersing agent for the CA. The resulting hydrated material is a composite of chemically bonded ceramic material and a cross-linked polyacrylate polymer.

The material as a paste or as a cured material was evaluated according to the tests in Table I. The material is either mixed by hand using a spatula by bringing the required amount of powder and liquid onto a mixing pad and mixing them thoroughly for 40 seconds, or by means of a capsule system. In the later case the powder and liquid have been pre-filled, in correct amounts to generate the required P:L ratio, into a dental capsule system. The capsule is first activated by bringing the powder and liquid together. The capsule is then transferred to a capsule mixing machine and mixed for a sufficient period of time. Using a 3M/ESPE Rotomix the time should be 8 s with a 3 s centrifuge stage in the end. After mixing the ready material is dispensed using a therefore suited tool, into any desired sample mould or container. There is no significant difference in properties depending on whether the material is mixed by hand or using a capsule system.

Table I: Standard tests used for evaluation of the dental luting cement

Test	Controlling standard
Net setting time	ISO 9917:2003 part 1
Film thickness	ISO 9917:2003 part 1
Compressive strength	ISO 9917:2003 part 1
Acid erosion	ISO 9917:2003 part 1
Radio Opacity	ISO 9917:2003 part 2
In vitro bioactivity	By testing apatite formation in phosphate solution

Microstructure and phase analyses have been evaluated by means of energy dispersive spectroscopy (EDS), scanning electron microscopy (SEM), transmission electron microscopy (TEM) and high-resolution TEM, and X-ray diffraction (XRD) including grazing incidence X-ray diffraction (GI-XRD) [6,7].

RESULTS

Basic properties for a dental luting agent

Setting time and film thickness
The most basic properties of a dental cement are working time, setting time, film thickness and mechanical strength. Working time for Ceramir C&B is two minutes, which is relatively long, while it has a setting time of approximately five minutes. Furthermore, Ceramir C&B has a low film thickness of around 15μm, which is a prerequisite for restorations having a good fit [8].

Compression Strength and Elasticity
Mechanical strength has been measured in terms of compression strength and gives 160 MPa after 24 hours, comfortably on par with the best resin-based materials, Table II.

Table II: Comparison of early strength (24 hrs) of some dental luting cement [9].

Material	RelyX Luting Cement	Fuji Plus	RelyX Unicem	Ceramir C&B
Compressive strength (MPa)	96 ± 10	138 ± 15	157 ± 10	160 ± 27

The compression strength of Ceramir C&B increases over time, and after approximately one month the strength is stabilised at 200 MPa (Table III).

Table III: Strength development over time for Ceramir C&B [9].

Time	24 hrs	8 days	30 days	90 days
Compressive strength (MPa)	160 ± 27	176± 24	196 ± 18	210 ± 24

Ceramir C&B's modulus of elasticity was measured at 4.7 GPa in combination with its compression strength evaluation.

Dimensional stability

Dimensional stability is a very important parameter for all dental cement. Ceramir C&B is a non-shrinking cement with the expansion during hardening close to zero, which is regarded as a great benefit since this precludes any great potentially harmful stresses between tooth and material. Even in tests where a "worst case" measurement case was sought the expansion is as low as 0.4 %. The contribution to the expansion in this case is free ongrowth of hydrated crystals from the surrounding liquid. The bulk expansion is zero. A thorough discussion on the topic is presented elsewhere [10,11].

To achieve more clinically relevant results, other tests were employed. One example of these was as follows: Entirely ceramic feldspar porcelain crowns were cemented with Ceramir C&B. These were evaluated over a period of time by passing light through them to test for micro-cracks. The tests were made on a number of occasions over 45 weeks and no cracks could be detected. Yet another test was carried out to get a picture of what pressure the material exerts on its surroundings. The material was placed in glass tubes (5, 7, 9 and 10 mm in diameter) with well-defined resistance against internal pressure. The largest tubes tolerated the least force of expansion. The tests were evaluated with a focus on crack formation over time. The study compared Ceramir C&B with Dyract Cem Plus (Dentsply) and Fuji Plus (GC). None of the glass tubes were broken by Ceramir C&B, whilst the other two cements cracked four and three tubes, respectively.

The results are shown in Table IV below.

Table IV: Glass tube test of some dental luting cements (numbers of broken glass tubes with time).

Material	At 2 weeks	At 3 weeks	At 4 weeks	At 19 weeks	At 5 weeks
Ceramir	0	0	0	0	0
Dyract Cem Plus	1 (5 mm)	2 (7 and 9 mm)	0	1 (10mm)	0
Fuji Plus	0	3 (7,9 and 10 mm)	0	0	0

Microstructure

One important property contributing to the microstructure developed by the CA-based dental luting material is that the calcium aluminates system during the hardening process has a high turn over of water. At temperatures above 30 °C the reaction is summarized as: a) dissolution of Ca-aluminate into the liquid, b) formation of ions, and c) repeated precipitation of nanocrystals – katoite, $3CaOAl_2O_36H_2O$ (C_3AH6), and gibbsite $Al(OH)_3$ (AH_3).

The main reaction for the mono Ca-aluminate phase is shown below (H=H$_2$O):

$$3CA + 12H \rightarrow C_3AH_6 + 2AH_3 \qquad (1)$$

The main reaction involves precipitation on contact areas and within the material, and repeated precipitation occurs until the Ca-aluminate or the water is consumed, resulting in cavity/gap/void/ filling. This makes it possible to reduce final porosity. This large volume of bonded water makes for great strength – many times greater than for normal calcium phosphates, which show a high porosity. In addition to this, the high water content gives for greater variation in its viscosity and composition, making it possible to use calcium aluminates in a broader spectrum of applications [12-15]. The general microstructure of the dental luting bioceramic material is shown in Fig. 3.

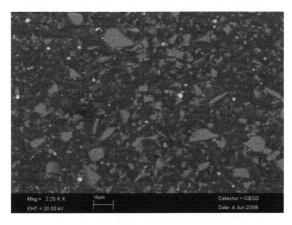

Fig. 1. Microstructure of the Ceramir^{TM} C&B material (bar = 10 μm)

The Figure 2 shows the nano-size structure of the hydrated phases.

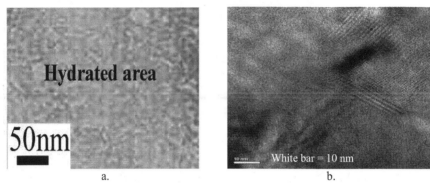

a.　　　　　　　　　　　　　　　　　b.

Fig. 2.　The nano-size structure of hydrated CA-phases, a) by TEM, and b) by HRTEM

The system exhibits a clear biocompatibility profile including showing bioactivity in contact with body liquid, i.e. a capacity to form hydroxyapatite in phosphate solution. Regarding the bioactivity, it has been shown by means of energy dispersive spectroscopy (EDS), scanning electron microscopy (SEM), transmission electron microscopy (TEM), grazing incidence X-ray diffraction (GI-XRD) that a layer of crystallised hydroxyl apatite is formed on the surface of the material when submerged in phosphate buffer solution [16-18]. See Fig. 3 below.

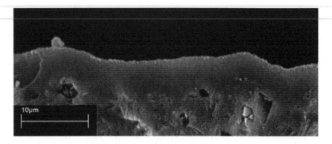

Fig. 3. Apatite formation on the surface of CA-based material.

Testing of Sealing

Microleakage

　　Utilisation of nanotechnology and nanostructural integration makes it possible to keep leakage between tooth and material to a minimum level over time. Leakage is normally measured by means of varying tests of micro- and nano-leakage. In these tests, differing kinds of colouration is used together with thermocycling to produce artificial aging of the joint. The material has been tested in both micro-leakage and bacterial-leakage models.

As a reference for the micro-leakage tests, a glass ionomer, Ketac Cem (3M ESPE) was used. Glass ionomers are the material group on the market which hitherto have generally displayed the lowest leakage factors. In the study, gold crowns were cemented to teeth which had been stored for 24 hours in a phosphate buffer at 37°C. After that, one group was thermocycled 2000 times from 5 to 55°C whilst the second group was placed directly into methylene blue dye. The teeth were then sectioned and each segment was then studied for leakage at each edge surface and evaluated. The test showed a significantly better seal of the surface edge when Ceramir C&B was used compared with Ketac Cem, see Table V. Data is in agreement with ealier published results of a Ca-aluminate material [19].

Table V: Micro-leakage data (Scale used in the table: 0°= zero leakage, 1° = 0-1 mm from crown cement surface, 2°= 1-2 mm, etc. [20])

Test Group	Medial leakage (grades) (Ceramir C&B n =120, Ketac Cem n =60)
Ceramir C&B after 24 hours	0,06
Ceramir C&B thermocycled	0,69
Ketac Cem after 24 hours	0,58
Ketac Cem thermocycled	1,22

As a complement to the micro-leakage tests, bacteria leakage evaluation was also carried out in a modified model designed for the testing of root-filling material. In this case, the control materials employed were a glass ionomer (Ketac Cem) and a resin-modified glass ionomer (RelyX Luting, 3M ESPE). After 60 days, the results of the study showed that both Ceramir C&B and RelyX Luting displayed generally low leakage and that both had significantly lower levels compared to Ketac Cem [20, 21].

Retention and shear bond strength
Retention has been tested by cementing crowns with an exaggerated convergence angle (32°). The purpose of the test was to measure the cement's retention ability without assistance of the preparation. The results showed that Ceramir C&B has a retention in parity with, or better than, RelyX Unicem against both gold and zirconium dioxide (Cercon, Degudent). Table VI summarises data for some dental luting cements with regard to the retention between the cements and the different materials [22].

Table VI: Retention data for some dental luting cements

Cement	Retention (gold crown) kg/force	Retention (zirconium dioxide crowns) kg/force
MaxCem (Kerr)	15.9 ± 9.3	nd
Ketac Cem	26.6 ± 4.4	nd
Zinc phosphate	13.9 ± 4.5	nd
Ceramir C&B	38.6 ± 8.5	32.6 + 6.7
RelyX Unicem	39.8 ± 15.3	27.8 + 11.3

Shear bond strength developed for Ceramir in contact with various other materials was compared with the performance of Ketac Cem. The Ceramir C&B material exhibited a high even shear strength independent of the contact surface. See Table VII.

Table VII: Shear bond strength in MPa towards different dental tissue and dental implant material for Ceramir and Ketac Cem dental luting cements [23]

Shear Bond Strength*	Ceramir™ C&B	Ketac Cem
Dentine	11.0	4.7
Enamel	8.4	8.4
Gold	10.2	2.8
Aluminium oxide	7.5	6.6
Zirconium dioxide	8.2	3.7

* Standard deviation in all tests was approximately 2 MPa

Complementary properties

Ceramir C&B fulfils the requirements of ISO 9917-1:2007 in terms of acidity-resistance after 24 hours. No detectable acid erosion was measured. A radio-opacity corresponding to 1.5 mm Al was established. A partial translucency has also been detected. Other important properties relate to thermal characteristics. The table below shows thermal characteristics of tooth structures and some dental luting cements.

Table VIII: Thermal properties [24]

Property	Dentine	Enamel	Ceramir materials	Glass ionomers	Composites
Thermal conductivity W/mK	0.57	0.93	0.8-0.9	0.51-0.72	1.1-1.4
Thermal expansion ppm/K	8.3	11.4	10	11	14-50

DISCUSSION

Below is discussed in some detail the general situation for dental luting cements, and an attempt is presented to summarise the status of available material classes used as dental cements.

Being able to create a completely impenetrable joint between dental material and the tissue of the tooth would be of great clinical benefit. The basic problem, however, is that after a while, a gap often occurs between the material and the dental tissue where nutrients and caries-producing bacteria can leak in. When the bacteria begin to metabolise the nutrient, caries develops and the dental tissue is destroyed. This phenomenon is quite common and is normally referred to as secondary caries. It can occur in all types of restorative work and is the main cause of all rectification work. The long-term consequence of secondary caries in prosthetic restorations is that the prosthetic work is eventually useless, either because of a failure in retention or because the tooth requires a root filling or needs to be extracted.

The reasons for the occurrence of gaps can be many, depending upon the type of material used. Certain older types of dental cement are recognised to have inadequate resistance to leaching in the mouth. This means that over time, material is gradually lost from the cement margin and this may finally produce leakage. Loss of material from the cement margin can also result from poor wear resistance, and this leads to material disappearing as a direct effect of wear over time.

A general reason for the formation of gaps, which applies to most material types in varying degrees, is that the material responds differently to the dental tissue when it is exposed to temperature changes. The environment of the mouth is one where substantial and rapid temperature changes can take place, and when that happens, everything expands or contracts. If there is a great difference between the thermal movements of the material and that of the dental tissue, stresses can eventuate, and these stresses can create a gap between material and tooth. This will be especially evident when general shrinkage during hardening/polymerisation occurs as for resin based materials.

To ensure a tight contact between material and dental tissue, the material needs to be able to initially wet the tooth effectively. Certain materials are hydrophilic and wet the water-retaining tooth naturally. Other materials are hydrophobic and require pre-preparation with multifunctional substances which are both hydrophilic and hydrophobic, whereby the hydrophilic constituent attaches directly to the tooth, and the hydrophobic one to the dental material.

A prerequisite against the incidence of leakage is that the material used in the restoration process is handled optimally. The technique sensitivity varies with differing materials but in general, resin-based materials require most from the clinician. In most cases, pre treatments will be necessary and the material will need to be bonded to the dental tissue so that it adheres. During hardening, shrinkage-initiated stresses take place and if the procedure has not been optimal, these in their turn produce gaps between material and tooth. The viscosity and the general rheology of the material are also important for a good contact. When the contact is created the material also has to attach to the dental tissue. Differing materials attach using differing mechanisms. Some of the most common attachment mechanisms employ mechanical locking between the tooth's irregularities and the material, chemical bonding and adhesion.

Resin-based materials are normally also sensitive to water and oxygen, which influence the hardening chemistry and/or the hardened material in a negative way. The least amount of fluid present during the period of hardening can eventually lead to the material leaving the tooth and/or becoming discoloured.

As shown in the result section above the new bioceramic dental luting cement Ceramir[TM] C&B has a promising general property profile with regard to general strength, thermal properties, retention, biocompatibility and stability [25, 26, 27].

The handling properties including working and setting time and film thickness and ease of removal of excess material are confirmed in a recent study comprising 500 treatments.

Also a one-year clinical study recently published [28] indicates the material to be an alternative to the best dental luting cements on the market. The 12 month recall data show that none of the fifteen patients reported any tooth/tissue sensitivity and that for all the restorations both the marginal integrity and discoloration criteria received "Excellent" scores. Furthermore, no caries was noted in association with any of the examined restorations [29].

In Table IX are presented the general features of all the dental cement classes available.

Tabel IX: Overview of dental luting cements.

Material aspects	Glass Ionomer	Resin-modified Glass Ionomer	Resin (bonded)	Self-adhesive Resin	Zn-phosphate cement	Ceramir C&B
Type of material	Polymer	Monomer	Monomer	Monomer	Inorganic material	Ceramic-polymer
Hardening mechanism	Cross-linking	Poly-merisation	Poly-merisation	Poly-merisation	Acid-base reaction	Acid-base + cross-linking
pH	Acidic	Acidic	Acidic/neutral	Acidic /neutral	Acidic	Acidic /basic
Geo-metrical stability	Non-shrinking	Non-shrinking	Shrinks	Shrinks	Non-shrinking	Non-shrinking
Stability over time	Degrades	Degrades	Degrades	Degrades	Degrades	Stable
Extra treatment	-	-	Etching and bonding	-	-	-
Hydro-philic / phobic	Hydro-philic	Hydrophilic	Hydrophobic	Initially Hydro-philic, Hydro-phobic	Hydrophilic	Hydro-philic
Integration mechanism	Micro-mechani-cal retention, Chemical bonding	Micro-mechanical retention / Chemical bonding / Adhesion	Adhesion / Micro-mechanical retention	Adhesion / Micro-mechanical retention	Micro-mechanical retention	Nano-structural integration
General behaviour	Irritant	Allergenic	Allergenic	Allergenic	Non-allergenic	Non-allergenic
Biocom-patibility	Good	OK	OK	OK	Good	Good
Bioactivty	No	No	No	No	No	Bioactive
Sealing quality	OK	OK	Good but operation sensitive	OK	Acceptable	Excellent

ACKNOWLEDGEMENT
The authors thank the personnel at Doxa AB and the professors Steven Jefferies and Cornelius Pameijer for valuable input.

REFERENCES
[1] Mjör et. al: Reasons for replacement of restorations in permanent teeth in general dental practice, International Dental Journal (2000) 50, 361-366
[2] Pjetursson B.E., Lang N.P.: Prosthetic treatment planning on the basis of scientific evidence, J. Oral Rehabilitation, 35 (Suppl. 1) 2008; 72-79
[3] Pjetursson B.E., Tan W.C., Tan K., Brägger U., Zwahlen M., Lang N.P.:
A systematic review of the survival and complications rates of resin-bonded bridges after an observation period of at least 5 years. Clin. Oral Impl. Res. 19, 2008; 131-141
[4] Jung R.E., Pjetursson B.E., Glauser R., Zembic A., Zwahlen M., Lang N.P.:
A systematic review of the 5-year survival and complication rates of implant-supported single crowns. Clin. Oral Impl. Res. 19, 2008; 119-130
[5] Pjetursson B.E., Brägger U., Lang N.P., Zwahlen M.: Comparison of survival and complication rates of tooth-supported fixed dental prostheses (FDPs) and implant-supported PDFs and single crowns (SCs). Clin. Oral Impl. Res. 18, (Suppl. 3), 2007; 97-113
[6] H. Engqvist, G. A. Botton, M. Couillard, S. Mohammadi, J. Malmström, L. Emanuelsson, L. Hermansson, M. W. Phaneuf, P. Thomsen: A novel tool for high-resolution transmission electron microscopy of intact interfaces between bone and metallic implants, Journal of Biomedical Materials Research, 78A (2006), 20-24
[7] H.Engqvist, T. Jarmar, Svahn, L.Hermansson, P. Thomsen: Characterization of the tissue-bioceramic interface in vivo using new preparation and analytical tools, Advances in Science and Technology, Vol 49 (2006), 275-281
[8] Technical data, Doxa Dental AB, 2009
[9] S.R. Jefferies, J. Lööf, C.H. Pameijer, D. Boston, C. Galbraith, L. Hermansson: Physical Properties of XeraCem™, J. Dental Res 87 (B), 3100, 2008
[10] L. Kraft, L. Hermansson, Gunilla Gómez-Ortega, A method for the examination of geometrical changes in cement paste: International RILEM 17, (2000) 401-4013
[11] L Kraft, L. Hermansson, Deformation characteristics in various CA admixtures with three different methods: Paper 3 in Ph D Thesis, L Kraft, Uppsala University 2002.
[12] Leif Hermansson, Adam Faris, Gunilla Gomez-Ortega, John Kuoppala and Jesper Lööf: Aspects of Dental Applications Based on Materials of the System, Ceramic Engineering and Science Proceedings, Volume 30, Issue 6, 71-80
[13] L. Hermansson, L. Kraft and H. Engqvist, Chemically bonded ceramics as biomaterials: Key Engineering Materials Vol. 247 (2003), 437-442
[14] L. Hermansson, H. Engqvist, J. Lööf, G. Gómez-Ortega, K.Björklund:
Nano-size biomaterials based on Ca-aluminate, Advances in Science and Technology, Vol. 49 (2006), 21-26
[15] L.Hermansson, J Lööf, T. Jarmar: Integration mechanisms towards hard tissue, Key Eng. Sci. Vol 396, 183 (2009)

[16] H. Engqvist, J-E. Schultz-Walz, J. Loof, G. A. Botton, D. Mayer, M.W. Phaneuf, N-O. Ahnfelt, L. Hermansson: Chemical and biological integration of a mouldable bioactive ceramic material capable of forming apatite in vivo in teeth. Biomaterials vol 25 (2004) pp. (2781-2787)

[17] J Lööf, F Svahn, T Jarmar, H Engqvist, C H Pameijer: A comparative study of the bioactivity of three materials for dental applications, Dental Materials Vol 24 (5), 653-659 (2008)

[18] L. Hermansson and H. Engqvist: Formation of apatite coatings on chemically bonded ceramics: Ceramic Transactions Vol 172 (2006) 199-206

[19] H. Engqvist, E. Abrahamsson, J. Lööf, L. Hermansson, Microleakage of a dental restorative material based on biominerals: Proceeding 29th International Cocoa Beach Conference and Exposition on Advanced Ceramics & Composites (January 2005)

[20] C.H. Pameijer, S. Jefferies, J. Lööf, L. Hermansson, Microleakage Evaluation of XeraCem™ in Cemented Crowns: J. Dent. Res. 87(B), 3098, 2008

[21] C.H. Pameijer, O. Zmener, F. Garcia-Godoy, J.S. Alvarez-Serrano: Sealing of XeraCem®, and controls using a bacterial leakage model, J. Dent. Res. 88(A), 3145, 2009

[22] C.H. Pameijer, S.R. Jefferies, J. Lööf, L. Hermansson: A comparative crown retention test using XeraCem™ , J. Dent. Res. 87(B), 3099, 2008.

[23] Physical properties of XeraCem, Shear bond strength of XeraCem, Data on file, 2008, Doxa Dental AB

[24] Phillips R.W. Science of Dental Materials. W. B Saunders Co. 1982, 8th ed.

[25] J. Lööf, Calcium-Aluminate as Biomaterial: Synthesis, Design and Evaluation. PhD Thesis Faculty of Science and Technology, Uppsala, University, Sweden (2008)

[26] L. Kraft, Calcium Aluminate Based Cements as Dental Restorative Materials, PhD Thesis Faculty of Science and Technology, Uppsala, University, Sweden 2002

[27] L Hermansson, A Faris, G Gómez-Ortega, J Lööf, Biocompatibility aspects of inject able chemically bonded ceramics of the system C-A-P-S: Ceramic Engineering and Science Proceedings, Volume 30, Issue 6, 59-70

[28] C.H. Pameijer, S.R. Jefferies, J. Lööf, L. Hermansson, E. Wiksell, In vitro and in vivo biocompatibility tests of XeraCem™: J. Dent. Res. 87(B), 3097, 2008

[29] S. Jefferies, C.H. Pameijer, D. Appleby, and D. Boston: One Month and Six Month Clinical Performance of XeraCem™, J. Dent. Res. 88(A), 3146, 2009

[30] S. Jefferies, C.H. Pameijer, D. Appleby, J Lööf and D. Boston: One-year Clinical Performance of XeraCem™, Accepted for publ in Swedish Dental Journal 2009.

BIOGLASS/CHITOSAN COMPOSITE AS A NEW BONE SUBSTITUTE

P. Khoshakhlagh[1], F. Moztarzadeh[1], S. M. Rabiee[2], R.Moradi[3], P.Heidari[4], R. Ravarian[1], S. Amanpour[4]

1. Department of Biomedical Engineering, Amirkabir University of technology, Hafiz Ave., Tehran, Iran
2. Oil Hospital, Hafiz Ave., Tehran, Iran
3. Department of Mechanic Enigineering, Babol (Noshirvani) University of Technology, Babol, Iran
4. Imam Khomeini Hospital, Keshavarz Blvd., Tehran, Iran

ABSTRACT

Glasses and glass-ceramic materials based on the SiO_2-CaO-P_2O_5 system constitute an important group of materials that have found wide application in medicine as bone implants. These materials are able to bind with bone in a living organism through the formation of apatite like layer on the implant site or surface. Sol-gel processing has been successfully used in the production of a variety of materials for both biomedical and nonbiomedical applications. The room temperature synthesis of single or multicomponent oxide glasses, in which the morphological, physical, and chemical properties can be tailored by synthesis parameters, is a significant advantage of this method. Chitosan is a biocopolymer comprising of glucosamine and N-acetylglucosamine, obtained by deacetylation of chitin. It has been reported to be safe, hemostatic and osteoconductive, and to promote wound healing. New injectable, bioglass/chitosan composites was investigated in this study. The prepared composite was characterized by Fourier transform infra-red spectroscopy (FTIR), X-ray diffraction analysis (XRD) and scanning electron microscopy (SEM). In-vivo evaluation was performed to examine the osteoconductivity of synthesized composite. In this case radiographic and pathological assays were accomplished. The objective of this study was to develop a new composite and the effect of chitosan on injectability was investigated.

INTRODUCTION

Bioactive glasses and glass-ceramics are used as implants to repair or replace parts of the body; long bones, vertebrae, joints, and teeth. Their clinical success is due to formation of a stable, mechanically strong interface with bone (Hench and Wilson 1993, Hulbert et al. 1987, and Hench 1998). Bioactive materials are typically made of compositions from the Na_2O-CaO, MgO-P_2O_5-SiO_2 system. The interest in bioactive materials is due to the fact that long-term ("20 years), clinical survivability of implants requires the formation of a stable interface with living host tissues. The mechanism of tissue attachment is directly related to the type of tissue response at the implant interface. No material implanted in living tissues is inert; all materials elicit a response from living tissues. Four types of response (Table I) allow different methods of achieving the attachment of prostheses to the skeleton. The relative chemical reactivity of different types of bioceramics correlates with the rate of formation of an interfacial bond of implants with bone. The relative level of reactivity of an implant also influences the thickness of the interfacial zone or layer between the material and tissue-implant interfacial strength. When biomaterials are nearly inert (Type 1 in Table I) and the interface is neither chemically nor biologically bonded, there is relative movement and progressive development of a nonadherent fibrous capsule occurs in both soft and hard tissues. Movement at the biomaterial-tissue interface eventually leads to deterioration in the function of the implant, or the tissue at the interface, or both. The thickness of the nonadherent capsule varies greatly, depending upon both material and the extent of relative motion.[1]

Table I. Types of bioceramics: tissue attachment and bioceramic classification.

Type of Bioceramic	Type of Attachment	Examples
1	Dense, nonporous, nearly inert ceramics attached by bone growth into surface irregularities by cementing the device into the tissues, or by press-fitting into a defect (Morphological Fixation).	Al_2O_3 ZrO_2
2	For porous implants, bone ingrowth occurs and mechanically attaches the bone to the material (Biological Fixation).	Porous hydroxyapatite
3	Surface-reactive ceramics, glasses, and glass-ceramics attached directly by chemical bonding with the bone (Bioactive Fixation).	Bioactive glasses Bioactive glass-ceramics Dense hydroxyapatite
4	Resorbable ceramics and glasses in bulk or powder form designed to be slowly replaced by bone.	Calcium sulfate Tricalcium phosphate

Results of in-vivo implantation show that these compositions produce no local or systemic toxicity, no inflammation, and no foreign-body response.[2,3] Recent research shows that there is genetic control of the cellular response of osteoblasts to bioactive glasses.[2,4] Compositions of sol–gel-derived bioactive glasses were used because they exhibit high specific area, high osteoconductive properties, and also a significant degradability.[2,5] Chitin is a co-polymer of N-acetyl-glucosamine and N-glucosamine units randomly or block distributed throughout the biopolymer chain depending on the processing method used to derive the biopolymer. When the number of N-acetyl-glucosamine units is higher than 50%, the biopolymer is termed chitin. Conversely, when the number of N-glucosamine units is higher, the term chitosan is used. Chitosan has been the better researched version of the biopolymer because of its ready solubility in dilute acids rendering chitosan more accessible for utilization and chemical reactions. In recent years, the biomaterials community has adopted this observation to generate several "composites" based on a polymer matrix–calcium-based compounds systems for potential use as hard tissue substitute materials. The advantage of such "composites" was believed to be the enhancement of the osteogenic potential accorded by the calcium compounds and the binder characteristic of the polymer matrix in inhibiting migration of the calcium compounds. For example, the combination of polymer with hydroxyapatite (HA) has been shown to in-vivo maximize the osteo-conductive behavior of HA, allowing bony ingrowths into the implant to occur as the matrix was progressively resorbed.[6,7,8] Therefore, HA or other calcium containing materials incorporated into chitin or chitosan has been a primary research area where orthopedic or bone substitution and periodontal applications were the focus. The injectability is important in clinical applications that involve defects with limited accessibility or narrow cavities, when there is a need for precise placement of the paste to conform to a defect area, or when using minimally invasive surgical techniques.[9]

Reduction of emerged heat which now is one of the main problems in materials which are being used as bone substitutes and reduction of emboli were some of the reasons that a new composition has been introduced, because the novel biomaterial is has two biodegradable components which are osteoconductive. Besides biodegradability, injectability is one of the most prominent purposes of proposing a new biocomposite. In addition, composite of two new components to enhance of the osteogenic potential bestowed by the calcium compounds and the binder characteristic of the chitosan considering the high osteoconductivity of bioglass was took into account. Therefore A new biomaterial as bone substitute was introduced.

MATERIALS AND METHODS

Synthesis of Bioglass

Bioglass was synthesized by sol- gel method. The composition of synthesized bioglass was based on CaO- SiO_2- P_2O_5 system (64% SiO_2, 31%CaO, and 5% P_2O_5), (based on mol %). The solution for the glass got ready as follows: 0.064 mol of Tetraethoxysilane (TEOS; Merck) was added into 30 mL of 0.1M Nitric acid; to complete the hydrolysis of TEOS, mixing process was prolonged for 30 minutes. 0.005 mol Triethylphosphate (TEP; Merck) and 0.031 mol Calcium Nitrate Tetrahydrate was added sequentially after 45 minutes and 90 minutes. Prepared sol isolated was cast in a Teflon container and kept sealed for 10 days at room temperature to let the hydrolysis and polycondensation reactions occur until the gel formation. The gel was maintained in the sealed container and heated at 70°C for a further 3 days. The water removal happened and a small hole was devised in the lid to allow the leakage of gases whilst heating the gel to 120°C for 3 days to get rid of the water. The parched gel was then heated for 24 h at 700°C for two reasons: the former to stabilize the glass and the latter to eliminate residual Nitrate.

Prepration of Bioglass/Chitosan Composite

The composite was prepared using a polymeric material that was chitosan regarding its effective properties for the final composite. Consequently Acetic acid (2wt %) was used as solvent. To achieve the best injectability and consolidation different solutions (2, 4, 6, 8, 10%wt) of chitosan in Acetic acid were prepared aged for 24hrs to get rid of bubbles. Afterward various compositions of bioglass in miscellaneous solutions of chitosan were examined and the most excellent composition from uttered standpoint was chosen (50%wt of bioglass in chitosan solution (2%wt))

Material Characterizations

X-ray Diffraction (XRD)

For X-ray diffraction analysis, the sintered samples were grinded and powdered. The resulting powders were characterized with Philips PW 3710 with 30 KV and 25 mA as its voltage and current respectively. Cu-Kα radiation was 1.540510 Å. For qualitative analysis, XRD diagrams were recorded in the interval $20° \leq 2\theta \leq 50°$ at scan speed of 2°/min.

Injectability Test

To assess the injectability of the composite a 1 ml syringe (diameter of 0.91 mm and head diameter of 0.2mm) was used. 1ml of the composite was filled into the syringe. The syringe was placed between two plates of compressive strength testing machine (Universal Testing Machine Zwick/Roell) and was extruded with the rate of 2mm/min. The force required to empty the syringe of composite was evaluated.

Scanning Electron Microscopy (SEM)

Composite was coated with a gold (Au) thin layer by sputtering (EMITECH K450X, England) and then the microstructure of the sample was studied on a scanning electron microscope (SEM; VEGA//TESCAN) with the acceleration voltage of 15 kV.

Fourier Transform Infrared Spectroscopy (FTIR)

The grinded composite was scrutinized by Fourier Transform Infrared Spectroscopy by the usage of Thermo Nicolet, USA. In the first step of this investigation, 1mg of the grinded sample was mingled with 300mg of KBr (infrared grade) and palletized under vacuum. Then the pellets were examined in the range of 400 to 4000 cm^{-1} at the scan speed of 23 scan/min with 4 cm^{-1} resolution.

Biological Evaluations:
In-vivo Test:
　　　To examine the osteoconductivity of the synthesized composite the sample was assessed through in-vivo test. 23 Adult male Sprague Dawley rats weighing approximately 250 g were used as the animal model. Surgery was performed under GA with Ketamine through IP injection .With longitudinal incision on the back of the rats we dissected the muscle and injected the sample between two transverse process of L4 and L5. Evaluation was conducted into three different time points (2, 4 and 8 weeks) for sham group(S) and material injected group (MI). In each time point, four animals of MI group and three animals of S were examined. Steps of surgical operation are depicted in Figure 1.

Radiographic Assay
　　　Rats were put under general anesthesia as explained, then they were fixed on radiography table and were exposed to X-ray (FLUOROSPOT, X-ray tube distance: 80 cm, window: 20×10 cm^2, 65 kV, 16.5 mev) and digital radiography was accomplished. Time point for radiographic assay was within 2, 4, and 8 weeks.

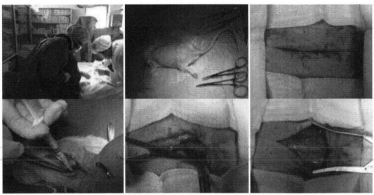

Figure 1. Operation steps of rats.

Pathological Assay
　　　After sacrificing the rats using IV injection of Nestonal, upper and downer vertebrae of the operation position were removed. Samples were dehydrated in 40% alcohol and in methylmethacrylate resin. In order to decalcification of samples, they were put in EDTA solution for two weeks. Then they were sagitally sected and surface stained with Toluidine blue.

RESULTS AND DISCUSSION
　　　It can be observed from injectability results (Figure 2) that there are two significant stages in its diagram. In the first stage a great deal of load is exerted to overwhelm the resistant forces against flow of the paste such as friction of sample and the body of the syringe. In Second stage the injection occurred on a stable rate. It was expected that chitosan behaves as a binder which results in a good injectablity of the composite.

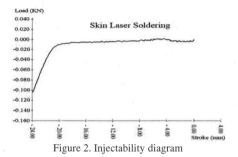

Figure 2. Injectability diagram

XRD analysis confirmed that after heat treatment the final product of sol-gel process was Bioglass and no distinct peak was observed. (Figure 3) it can be concluded that just a little amount of crystallization has been occurred during heat treatment which can be relinquished. From the XRD pattern it can be concluded that the produced bioglass is nearly amorphous.

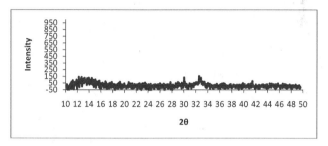

Figure 3. XRD pattern of synthesized bioglass

After immersion in Ringer's solution, adequate clues were detected which could prove the formation of HA. All P-O vibrations are evident in the range of 600-700 cm^{-1} after immersion. Peaks in the range of 1100 cm^{-1} can be imputed to Si-O-Si bond that seems to be expanded after immersion; it demonstrates the dissolution of these bonds which can be an overture to HA forming. (Figure 4)

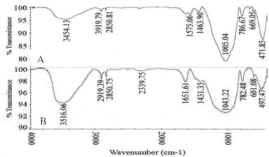

Figure 4. FTIR results of the sample
A) Before soaking in Ringer's solution B) After soaking in Ringer's solution

It can be concluded from the SEM images that chitosan acts as a binder for bioglass powders and that can be the cause for paste-like behavior. It can be detected that bioglass powders have been dispersed evenly in chitosan matrix and chitosan kept the particles by its cohesiveness. The key for good injectability of the composite is this characteristic of chitosan beside the small average size of bioglass particles.(Figure 5)

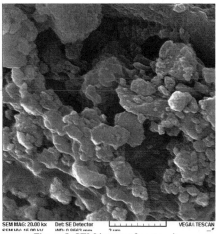

Figure 5. SEM image of composite

Mice were taken X –rays at 2 weeks, one month and 2 months. Radiographic images exhibited the gradually absorption of the material and at the end of the 2 months; it was disappeared completely in all rats. Fusion and new bone formation was observed not only at the surgical level but also at the following levels. (Figure 6)

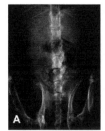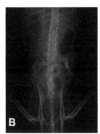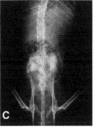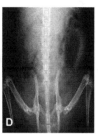

Figure 6. X-rays of rats
A) After 8 weeks, B) After 4 weeks, C) After 2 weeks, D) Sham after 8 weeks

Pathological assay showed that bone formation had been accelerated noticeably in comparison with sham samples. New bone formation was observed in surgical level of the rats where injection was carried out. Figure 7 depicts pathological images of bone formation process in rat sample during 2, 4 and 8 weeks and right after injection. In the first step of injection, the biocomposite can be demonstrated everywhere. After two weeks biocomposite is going to be absorbed and disappeared and granulation tissue was replaced gradually. After four weeks, cartilage formation can be detected in the left side of the figure 7-C however; osteoid formation was detected in the right side and in the middle. After four weeks the figure demonstrates cartilage formation in the left side, in the right side and middle of the field osteoid formation existed. During eight weeks of injection, pathological images showed normal bone formation for instance in the Figure 7-D normal bone can be seen in the right side.

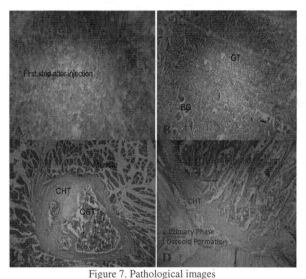

Figure 7. Pathological images
A) First step after injection, B) After two weeks, granulation tissue has been formed
C) After 4 weeks, cartilage has been formed, D) After 8 weeks, new bone formation can be observed

CONCLUSION

Injectability test demonstrated good injectable properties of Bioglass/Chitosan as a composite. There were some conclusions from SEM images 1) bioglass has been dispersed in polymer matrix 2) chitosan is a binder and it enhanced the osteogenic potential due to its binder characteristic3). FTIR illustrated that after incubation in Ringer's solution, HA has formed. Radiographic and Pathological assays approved that beside all other properties which was expected such as injectability, bone formation occurred even in the places that was not expected. It shows that the composite of bioglass/chitosan is highly osteoconductive.

REFERENCES:

[1]Encyclopedia of Materials: Science and Technology, *Elsevier Science Ltd*, 563-568 (2001).

[2]N. Li et al., Preparation and Characterization of Macroporous Sol–Gel Bioglass, *Ceramics International*, **31**, 641-646 (2005).

[3]P. Sepulveda, J.R. Jones and L.L. Hench, Bioactive Sol–Gel Foams for Tissue Repair, *J. Biomed. Mater. Res.*, **59** 340-348 (2002).

[4]L.L. Hench and J.M. Polak, Third-generation Biomedical Materials, *Science,* **295**, 1014-1017 (2002).

[5]P. Saravanapavan and L.L. Hench, Low-temperature Synthesis, Structure, and Bioactivity of Gel-derived Glasses in The Binary CaO–SiO2 System, *J. Biomed. Mater. Res.*, **54**, 608-618 (2001).

[6]E. Khor, L.Y. Lim, "Implantable Applications of Chitin and Chitosan", *Biomaterials*, **24**, 2339-2349 (2003).

[7]Higashi S, Yamamuro T, Nakamura T, Ikada Y, Hyon SH, Jamshidi K. Polymer-hydroxyapatite Composites for Biodegradable Bone Filler, *Biomaterials,* **7**, 183-7 (1986).

[8]Ito M. In vitro properties of a chitosan-bonded hydroxyapatite bone-filling paste. *Biomaterials,* **12**, 41-5 (1991).

[9]H.H.K. Xu et al., Injectable and Macroporous Calcium Phosphate CementS, *Biomaterials*, **27**, 4279-4287 (2006).

DEVELOPMENT AND CHARACTERIZATION OF HIGH STRENGTH POROUS TISSUE SCAFFOLDS

James J. Liu, Juha-Pekka Nuutinen, Kitu Patel and Adam Wallen

GEO2 Technologies
12-R Cabot Road
Woburn, MA 01801.

ABSTRACT

The aim of this study is to develop bioactive scaffolds for bone tissue engineering based on bioactive glass fiber with novel processing techniques. The fabrication of such novel scaffolds involves pore formation and strengthening sintering techniques which leads to a well defined cross linked microstructure with improved mechanical properties. A highly interconnected pore network is created by selection and arrangement of pore-formers and fibers. In addition, these three dimensional porous bioactive glass scaffolds exhibit better mechanical properties compared to non-strengthening scaffolds due to strengthening during sintering. The porosity can be controlled from 20 to 75% with pore size ranged from 50 to 500 micron. *In-vitro* reactions of these unique microstructure bioactive glass scaffolds are studied with aqueous solutions of simulated body fluid (SBF).

We have successfully developed porous bioactive glass scaffolds with improved mechanical properties and well controlled porosity and interconnected open pores. The potential applications of this high strength bioactive glass fiber-based scaffold are load-bearing applications, such as implants for spinal column and extremities.

INTRODUCTION

Prosthetic devices are often required for repairing defects in bone tissue in surgical procedures. Prostheses are often required for the replacement or repair of diseased or deteriorated bone tissue to enhance the body's own mechanism to produce rapid healing of musculoskeletal injuries resulting from severe trauma or degenerative disease. Various types of synthetic implants have been developed for tissue engineering applications in an attempt to provide a synthetic prosthetic device that mimics the properties of natural bone tissue and promotes healing and repair of tissue.[1-4]

The challenge in developing a resorbable tissue scaffold using biologically active and resorbable materials is to attain load bearing strength with porosity sufficient to promote the growth of bone tissue. Conventional bioactive bio-glass and bio-ceramic materials in a porous form are not known to be inherently strong enough to provide load bearing strength as a synthetic prosthesis or implant.[2,4] Similarly, conventional bioactive materials in a form that provides sufficient strength do not exhibit a pore structure that can be considered to be osteostimulative.

Fiber-based structures are generally known to provide inherently higher strength to weight ratios, given that the strength of an individual fiber can be significantly greater than powder-based or particle-based materials of the same composition. A fiber can be produced with relatively few discontinuities that contribute to the formation of stress concentrations for failure propagation. By contrast, a powder-based or particle-based material requires the formation of bonds between each of the adjoining particles, with each bond interface potentially creating a stress concentration.

Furthermore, a fiber-based structure provides for stress relief and thus, greater strength, when the fiber-based structure is subjected to strain in that the failure of any one individual fiber does not propagate through adjacent fibers. Accordingly, a fiber-based structure exhibits superior mechanical strength properties over an equivalent size and porosity than a powder-based material of the same composition.[7-8]

In this work we demonstrate a family of novel, microstructurally-ordered materials – the fibrous cross-linked microstructure (CLM) ceramic materials. Bio-ceramics produced with the fibrous CLM ceramic materials possess the advantages of a high porosity macro-structure and the high performance mechanical properties of the fibrous microstructure. There are significant differences between fibrous CLM materials and fibrous materials, such as those with sintered fiber structures, documented in previous literature[9-16]. The aim of this study is to characterize the fibrous CLM bio-ceramics and their control performance advantages for bone tissue engineering applications.

EXPERIMENTAL PROCEDURES

A CLM technique will be utilized to produce test samples with different porosities and pore interconnectivity. The CLM process will be used to produce scaffolds with a range of different porosities, pore sizes, mechanical properties and *in-vitro* degradation rates. The diameter and composition of the fibers, along with the size and raw material of the chosen pore-formers and binders, will influence the resulting pore size and pore size distribution of the structure. The influence of the mentioned process parameters on the scaffold properties will be discussed. The fibrous CLM process includes a fibrous and powder mixture extrusion process. Fibrous CLM bio-ceramics were extruded from ceramic paste which was prepared by mixing >50 wt% of fiber with addition of binders, and pore formers. In an example of a CLM bio-glass material, 13-93 glass fiber was used as a raw material and mixed with pore-formers and sintering aids. 13-93 glass, a glass composition having a composition in respective mol% quantities of 6% Na_2O; 7.9% K_2O; 7.7% MgO; 22.1% CaO; 0% B_2O_3; 1.7% P_2O_5; and 54.6% SiO_2, were mixed and drawn using an in-house fiber drawing equipment. Both process yield and fiber consistency were evaluated as the fiber composition, diameter and process parameters altered. Starting fibers were chopped to size with length range from 0.5 cm to 2cm and diameter greater than 15 micron. These constituents were mixed in a high-shear sigma-blade mixer for ~60 minutes until a ceramic paste was formed. During mixing fibers were chopped to more uniform length. The pastes were then de-aired, formed into a billet and extruded into a 12.5 mm round rods using a stainless die in a hydraulic ram extruder.

The extruded fibrous CLM bio-ceramics rods were analyzed for porosity and compressive strength measurements. Porosity of the substrate was measured using both standard dimension and weight and mercury intrusion porosimetry techniques (Micromeritics, AutoPore IV 9500 V1.07). Bars were also cut out from the sintered parts to measure the Young's modulus and modulus of rupture, the rods or bars were roughly 5 mm in diameter by 10 mm in length.

The compressive strength, elastic modulus under compression, and respective failure modes were determined, since those properties are critically important in several orthopedic bone scaffolding applications. A standard test protocol, for example, ASTM F451-08, was modified to fit the bioactive glass scaffold testing. The axial compressive strength was measured from cored samples on an Admet universal testing machine with a crosshead speed of 22.2 N/second. Compressive strength was measured in the direction parallel to the axis of extrusion. The microstructure was analyzed using scanning electron microscopy performed at Alfred University. The operating voltage used was 20 kV.

A part of the proof of concept testing included *in-vitro* simulated body fluid (SBF) testing to validate the bioactivity of the scaffold. The porous bodies were immersed in simulated body fluid (SBF) to evaluate their degradation process. The composition of the immersion fluid and the method to produce it were based on the work of Kokubo et al. [17,18] Previous studies have shown that the parameters affecting the *in-vitro* degradation process of bioactive materials are numerous, in particular, immersion temperature (T), time (t) and surface-area-to-volume ratio (SA/V).[17-19]

Samples of the raw material fiber and scaffolds were placed in polystyrene bottles containing simulated body fluid (SBF) with ion concentrations nearly equal to human blood plasma.[17,18] The bottles with the samples and SBF were maintained at 36.5 °C in a shaking water bath for 1, 7, and 14 days respectively without refreshing the soaking medium. The sample surface area to SBF volume (SA/V) ratio of 0.1 cm^{-1} was used for all the test samples. After various soaking periods, the samples were filtrated and gently rinsed twice with Ethanol to remove SBF followed by drying in vacuum at 80 °C. Three sets of analysis were performed for samples with in vitro conditioning, namely: (1) changes in pH of the solutions; (2) change in mass; and (3) scanning electron microscopic (SEM) analysis and compositional analysis of the glass surface after immersion in SBF. The formation of HAp on the surface of the fibers and scaffolds was characterized by XRD the morphology of the samples was observed using SEM.

RESULTS AND DISCUSSION

Cross-linked microstructure (CLM) materials in various compositions have been developed for a number of applications, including filtration and catalytic hosts. For example, porous ceramic bodies are of particular use in the automotive industry as honeycomb substrates to host catalytic oxidation and reduction of exhaust gases. The CLM material in a ceramic honeycomb forms provide much higher strength and the requisite high specific surface area for filtration and support for catalytic reactions in these applications.

It is demonstrated that several specific bio-ceramics (including bio-active glass, bio-inert alumina, and silicon carbide) can be extruded into novel CLM scaffolds. Macro-structure of scaffolds is controlled by the die design and extrusion parameters. Three different types of extrusion were used in this study, including honeycomb extrusion, co-extrusion, and single rod extrusion, as shown in Figure 1. Honeycomb structure scaffold is extruded with honeycomb die with continuous channels and divided walls. Honeycomb extrusion is more complex but it provides much higher porosity and complexity geometry as shown in Figure 1b. The other potential benefits of extrusion techniques include low cost processing, vast materials property design and applications.

The CLM microstructure produces a combination of high strength at high porosity and high material permeability for a given pore size distribution. The unique material properties of CLM scaffolds are due to the cross-links between fibers and the crystal and grain structure induced through chemistry and sintering. Fibers are arranged in a 3-D interconnected matrix during mixing and extrusion. Fibers are cross-linked through sintering and *in situ* chemical reactions. The sintering process is necessary to obtain a mechanically rigid three-dimensional CLM substrate. The strength of the substrate is derived from the fibers and bonds between overlapping and adjoining fibers within the structure. Scanning electron micrographs (SEM) of CLM scaffolds are shown in Figure 2. These images indicate that the cross-linked microstructure ranges from fibrous bio-active glass fiber to fine crystalline alumina fiber. The cross-linked microstructure can be provided in an extruded textured pattern that is formed from ceramic fibers. The resulting cross-linked microstructure has higher connection points for additional crossed-linked bonding. For the extruded honeycomb scaffold, it has an additional array of opening channels forming walls between adjacent channels, with the wall formed from intertangled fibers. Anisotropic results should not be surprising for extrusion process given the inherent structure of fibers itself, the unusual textured structure, and the complex chemistry of extruded fibrous CLM materials. In a micro-scale, the fibrous structure introduces a preferred-orientation microstructure, i.e. fibers are preferably aligned along the extrusion direction, due to the severe plastic characteristics of the paste. This type of CLM porous network may therefore prove useful as scaffold in applications where load bearing materials are required.

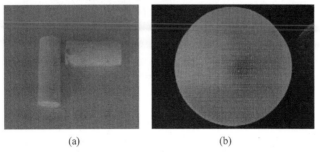

(a)   (b)

Figure 1. Extruded CLM Bio-Ceramics.(a) Single rod extrusion (b) Honeycomb extrusion.

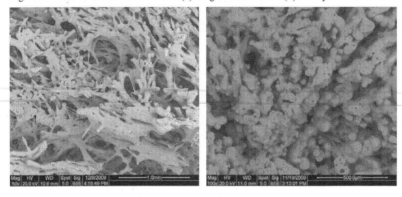

Figure 2. SEM of CLM Bio-glass.(Back Scattered Electron Image)

Porosity and pore size distribution are important parameters for the scaffold application. The advantage of high porosity in scaffolds has been reported to achieve good cellular distribution, and that it is critical to select the correct pore size.[12] Porosity can, however, adversely affect important mechanical characteristics of a scaffold, requiring more complex material designs. It is more difficult to achieve a high porosity with high strength than variation of micro-structure. This challenge may be overcome by re-engineering a porous scaffold using fibrous structure. In so doing, it is possible to extend the porosity range at the same time, while maintaining adequate mechanical strength. High porosity is one of the key characteristics of fibrous structured materials. The porosity of a fibrous structure is created by open space between fibers. A highly porous fibrous structure is also utilized in many other applications, such as fiber insulation and fiber filtration. The results indicate that extruded fibrous CLM ceramics have porosity greater than 50% and can reach up to 80% and maintain equivalent strength. The porosity of extruded fibrous CLM ceramics depends on several factors, such as fiber diameter, pore-former, chemical reactions and sintering. Figure 5 shows the porosity as function of sintering temperature of 13-93 bio-glass produced by CLM and traditional sintered fiber methods. It is indicated that the scaffolds made from CLM technique has higher porosity compared with sintered fiber.

The CLM pore structure is different from traditional particle-based or fiber based ceramics. Pore structure of sintered particle or fibers is limited by sintering conditions, such as temperature, and time. Pore structure of CLM ceramics can be controlled by chemistry and sintering. CLM in a 13-93 bio-glass composition exhibit a bimodal pore size distribution with the large pore size range from 50 micron to 350 micron. This is particularly important for applications that require load bearing application, where large pores or open channels are needed for vascularization. This novel fibrous CLM was also examined with three-dimensional visualization techniques. In addition, X-ray nano-CT was used to investigate the CLM scaffold and demonstrate its unique characteristics in 3-D at high resolution. Detailed quantitative results from the X-ray nano-CT analysis of several CLM ceramics will be published separately.

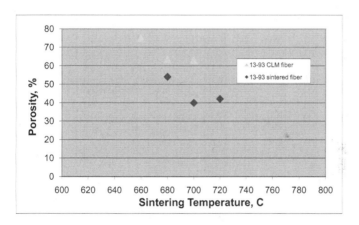

Figure 3 Porosity of the scaffolds as a function of sintering temperature for 13-93 bio-glass CLM and sintered fiber.

The process of bone graft incorporation is similar to the bone healing process that occurs in fractured long bones.[6] Bone grafts are also strongly influenced by local mechanical forces during the remodeling phase. Mechanical demands modulate the density, geometry, thickness and trabecular orientation of bone, allowing to optimize the structural strength of the graft. The fibrous CLM has much improved mechanical properties compared to particle-based microstructures, because of the inherent characteristic of fiber. The porosity effects on mechanical properties are also shown in Figures 4 and 5. They indicate that the fibrous CLM materials have adequate mechanical strength at high porosity. The relationship between mechanical properties and porosity can be analyzed with various analytical models. In preliminary results, there is a correlation between strength and porosity for CLM, and this correlation may be different from porous particle-based ceramics. Accurate prediction of the strength of porous CLM materials requires more adequate information for comparison. It is necessary to point out that the differential results of theoretical predication models may arise from irregular shapes, varied size and random distribution of the pores and others.

For similar porosity, CLM bio-ceramics have higher mechanical properties compare to sintered fiber-based bio-ceramics. The improved mechanical performance of CLM ceramics may be due to the unique microstructure and chemistry. The microstructural evolution of CLM ceramic materials with increasing porosity is related to 3-D connectivity. Connectivity represents the contact area between fibers or fibrous grains. The increasing connectivity results in increased mechanical strength as discussed in previous section, however, the bonding strength between the fibers is important for increased load bearing. The minimum contact solid area (load-bearing area) which is the actual sintered or the bond area is significantly different between particle-based and fiber-based materials.

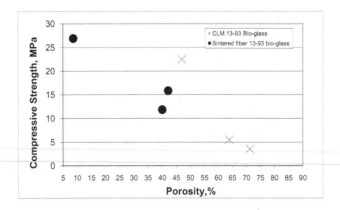

Figure 4. Comparison of compressive strength vs porosity of CLM 13-93 bio-glass and sintered fiber-based 13-93 bio-glass.

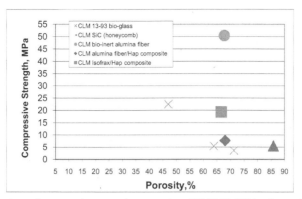

Figure 5. Comparison of compressive strength vs porosity of CLM 13-93 bio-glass and CLM bio-inert materials

Based on the promising CLM microstructures and the unique elastic plastic mechanical response, this study was undertaken to evaluate the in SBF *in-vitro* response to bioactive glass scaffolds prepared by the CLM extrusion process. Bioactive glass with the 13-93 composition was used because of its proven bioactivity,[10] as well as its ability to support cell proliferation and function,[11,12] The glass is also approved for *in-vivo* use in the United States and elsewhere. 13-93 bioactive glass scaffolds with the CLM microstructure had more favorable pore characteristics for tissue in-growth.

The results indicated that fibers, similar to the 13-93 powder,[16] could also develop bone-like HAp layers on its surface when exposed to SBF. It is a common notion that bone-like HAp, which is a synthetic calcium phosphate that resembles bone mineral, plays an essential role in the formation, growth and maintenance of the tissue-biomaterial interface.[19-22] Therefore, the present study suggests that scaffold made from CLM 13-93 bio-glass possess good bioactivity, and may be used for preparation of bioactive scaffolds. Fig. 6a and b shows SEM image of 13-93 CLM scaffold after soaking in SBF for 14 days. From Figure 6a, it can be seen that bone-like HAp forms on the surface. The EDAX spectra is shown in Figure 6c, which indicates that the characteristic peaks of SiO_2 disappeared and a new calcium HAp, were apparent after soaking for 14 days. After prolonged soaking for 7 and 14 days, HAp was detected in the XRD patterns (Figure 7). Continued biomechanical testing and *in-vivo* testing through a small animal model is in process.

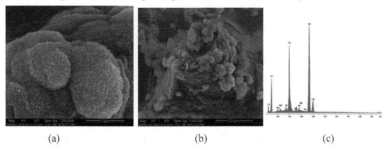

(a)   (b)   (c)

Figure. 6 HAp formation for CLM 13-93 bio-glass after immersion in SBF for 14 days.

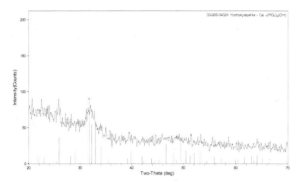

Figure. 7 XRD of HAp formation for CLM 13-93 bio-glass after immersion in SBF for 14 days

CONCLUSIONS

There is a demand for the development of improved biomaterials to be used as scaffolds in bone and cartilage tissue engineering. Bioactive glass fibrous scaffolds with cross-linked microstructure may be used to meet this demand, since bioactive glass fibers possess osteoconductivity, provide high strength and offer a possibility to develop new materials and manufacturing methods. Highly porous three-dimensional fibrous cross-linked microstructure network can be obtained. The porosity of a bioactive glass network not only noticeably increases the total reacting surface of the glass, but also serves as a framework for tissue in-growth. This type of material may therefore prove useful as scaffolds for tissue engineering applications, for filling bone defects after trauma, infection and surgery.

ACKNOWLEDGMENTS

We wish to thank contributing team members at GEO2 Technologies, including Art O'Dea, Leonard Newton, and Rob Lachenauer.

REFERENCES

1. D. W. Hutmacher, "Scaffolds in Tissue Engineering Bone and Cartilage," Biomaterials, 21, 2529–43 (2000).

2. L. L. Hench, "Bioceramics," J. Am. Ceram. Soc., 81, 1705−28 (1998).

3. M. M. Stevens, "Biomaterials for Bone Tissue Engineering," Mater. Today, 11, 18–25 (2008)

4. J. R. Jones, E. Gentleman and J. Polak, "Bioactive Glass Scaffolds for Bone Regeneration," Elements 3[6], 393-399 (2007

5. L. J. Bonassar and C.A. Vacanti, "Tissue Engineering: The First Decade and Beyond". J Cell Biochem. Suppl. 30–31:297–303 (1998).

6. M. A. Meyers, P. Chen, A. Y. Lin, and Y. Seki, "Biological Materials: Structure and Mechanical

Properties," Prog. Mater. Sci., 53, 1-206 (2008).

7. B. Zuberi, J. J. Liu, S. C. Pillai, J. G. Weinstein, A. G. Konstandopoulos, S. Lorentzou, and C. Pagoura, "Advanced High Porosity Ceramic Honeycomb Wall Flow Filters" SAE Paper 2008-01-0623 (2008).

8. J. J. Liu, R. A. Dahl, T. Gordon, and B. Zuberi, "Use Of Ceramic Microfibers To Generate A High Porosity Cross-linked Microstructure In Extruded Honeycombs," 33rd International Conference & Exposition on Advanced Ceramics & Composites, (2009)

9. M Brink, "Bioactive glasses with a Large Working Range" Doctoral thesis, Åbo Akademi University, Turku, Finland; 1997.

10. M. Brink, T. Turunen, R. Happonen, and A. Yli-Urppo, "Compositional Dependence of Bioactivity of Glasses in the System $Na_2O-K_2O-MgO-CaO-B_2O_3-P_2O_5-SiO_2$," J. Biomed. Mater. Res., 37, 114–21 (1997)

11. M. Brink, "The Influence of Alkali and Alkaline Earths on the Working Range for Bioactive Glasses," J Biomed. Mater. Res., 36:109–17 (1997).

12. E. Pirhonen, M. Loredana, and J. Haapanen, "Porous Bioactive 3-D Glass Fiber Scaffolds for Tissue Engineering Applications Manufactured by Sintering Technique," Bioceramics 15. (2002)

13. E. Pirhonen "Porous Bioactive 3-D Glass Fiber Scaffolds For Tissue Engineering Applications," Key Engineering Materials Vols. 240-242, 237-240 (2003).

14. R. F. Brown, D. E. Day, T. E. Day, S. Jung, M. N. Rahaman, and Q. Fu, "Growth and Differentiation of Osteoblastic Cells on 13–93 Bioactive Glass Fibers and Scaffolds," Acta Biomater., 4, 387–96 (2008).

15. Q. Fu, M. N. Rahaman, B. S. Bal, W. Huang, and D. E. Day, "Preparation and Bioactive Characteristics of a Porous 13–93 Glass, and Fabrication into the Articulating Surface of a Proximal Tibia," J. Biomed. Mater. Res., 82A, 222–9 (2007).

16. Q. Fu, M. N. Rahaman, B. S. Bal, R. F. Brown, and D. E. Day, "Mechanical and In Vitro Performance of 13-93 Bioactive Glass Scaffolds Prepared by a Polymer Foam Replication Technique," Acta Biomater., 4, 1854-64 (2008).

17. T. Kokubo, H. Kushitani, S. Sakka, T. Kitsugi, T. Yamamuro, "Solutions Able To Reproduce In Vivo Surface-Structure Changes in Bioactive Glass–Ceramic A–W," J. Biomed. Mater. Res. 24, 721–734 (1990).

18. T. Kokubo and H. Takadama, "How Useful Is SBF in Predicting In Vivo Bone Bioactivity?," Biomaterials, 27 ,2907–2915 (2006)

19. I. D. Xynos, M. J. Hukkanen, J. J. Batten, L. D. Buttery, L. L. Hench, J. M. Polak, "Bioglass 45S5 Stimulates Osteoblast Turnover and Enhances Bone Formation in Vitro. Implications and Applications For Bone Tissue Engineering" Calcif. Tissue Int. 67, 321–9 (2000).

20. L. L. Hench, R. J. Splinter, W. C. Allen, and T. K. Greenlee Jr., "Bonding Mechanisms at the Interface of Ceramic Prosthetic Materials," J. Biomed. Mater. Res., 2, 117–41 (1971).

21. M.A. De Diego, N. J. Coleman, and L.L. Hench, "Tensile Properties of Bioactive Fibres for Tissue Engineering Applications," J. Biomed. Mater. Res. (Appl. Biomater.), 53:199–203 (2000).

22. D. L. Wheeler, K. E. Stokes, R. G. Hoellrich, D. L. Chamberland, and S. W. McLoughlin, "Effect of Bioactive Glass Particle Size on Osseous Regeneration of Cancellous Defects," J. Biomed. Mater. Res., 41, 527–33 (1998).

GRAPE® TECHNOLOGY OR BONE-LIKE APATITE DEPOSITION IN NARROW GROOVES

A. Sugino, K. Uetsuki, K. Kuramoto
NAKASHIMA MEDICAL Co., LTD.
688-1 Joto-Kitagata, Higashi-ku, Okayama 709-0625, Japan

S. Hayakawa, Y. Shirosaki, A. Osaka
Graduate School of Natural Science and Technology
Research Center for Biomedical Engineering, Okayama University
Tsushima, Kita-ku, Okayama 700-8530, Japan

K. Tsuru
Faculty of Dental Science, Kyushu University
Maidashi, Higashi-ku, Fukuoka 812-8582, Japan

T. Nakano
Graduate School of Engineering, Osaka University
Yamada-oka, Suita 565-0871, Japan

C. Ohtsuki
Graduate School of Engineering, Nagoya University
Furo-cho, Chikusa-ku, Nagoya-shi 464-8603, Japan

ABSTRACT
Among several techniques proposed for fixing titanium bone implant in bone tissues by inducing apatite deposition on the implant surface, "*GRAPE® Technology*" is the simplest one, where optimum size of macro-grooving and heating in air are provided for pure titanium because samples with such grooves deposit apatite when soaked in an aqueous solution mimicking human blood plasma. The technique is needed to be confirmed by *in vivo* tests if it is effective for bone tissues, or if it is applicable to newly developed titanium alloy systems.
In this study, the validity of GRAPE® Technology for a Ti-15Zr-4Ta-4Nb alloy was assessed with both *in vitro* and *in vivo* experiments. Here, alloy pieces with macro-grooves 500 μm in depth and 500 μm in width were heated at 500 °C to induce thin titania layers on the surface. *In vivo* tests confirmed bone-ingrowth within 3 weeks in the samples with the macro-grooves but very less for those without the grooves. Those results indicated that GRAPE® Technology is effective for providing the Ti-15Zr-4Ta-4Nb-based implants with osteoconductivity.

INTRODUCTION
Secured fixation of bone-substitutes or bone implants with bone is essential in dental and orthopedic fields. Activity of spontaneously depositing apatite, similar in composition and structure involving lattice defects to bone tissue apatite, should be provided for implant surfaces for accomplishing fixation. Such ability is sometimes denoted as bioactivity, for which the implants are strongly integrated with bone tissues. Titanium and its alloys are marginally active in forming such direct bonds with human bone tissues though they are frequently used as bone-substitute materials. Several surface modification techniques have been proposed to provide such implants with the ability of bone-implant fixation; for example, titania layer coatings by chemical[1-3], thermal[4-6] or electro-chemical treatments[7, 8], bioactive ceramic coatings due to plasma spray[9] or electrophoresis[10]. Those techniques control the structure, composition, and chemistry of the titanium surfaces. In principle, they provide the surface layers active to induce spontaneous apatite deposition under body

environment. Yet, they have their own advantages and disadvantages. Amorphous or crystalline titania derived by those chemical treatments does not allow apatite deposition *in vitro* in a simulated body fluid (Kokubo's SBF[11]) within a short time unless it is heated at moderate temperatures[5,6] or deposited with calcium ions[8]. Active ceramic coatings involve difficult and complicated fabrication processes, and will be instable at the interface between the coated layers and substrates after long-term implantation.

Wang *et al.*[6] reported that rutile layers derived on pure titanium substrates by thermal oxidation at 400 °C were active: they set together two pieces of such specimens in a V-letter shape geometry with a 600 μm-mouth opening and soaked in SBF to find that apatite was deposited on both facing surfaces. They pointed out the importance of the narrow spaces confined by the two specimens. Following Wang *et al.*[6], Sugino *et al.*[12, 13] machined pure titanium specimens to derive macro-groove of 500 μm in depth and 200-1000 μm in width and heated in air at 400 °C. They also confirmed many apatite hemi-spherical particles on the groove walls and bottom. These results indicated that both appropriate specific spaces and thermal oxidation temperature could enhance the potential for apatite formation on titanium. The technology relevant to the spatial design above is named GRAPE® Technology (*i.e.*, GRoove and APatitE), and is under refinement for medical applications. Yet, the applicability should be confirmed on some titanium alloys since titanium alloys are frequently used for implants but not pure titanium. GRAPE® Technology is epoch-making because the implants are only heated in air and proper narrow spaces are prepared so that they will firmly bond to surrounding bone tissues.

In this study, the validity and applicability of GRAPE® Technology to a Ti-15Zr-4Ta-4Nb alloy was assessed with *in vitro* and *in vivo* experiments. Ti-15Zr-4Ta-4Nb has been reported to show much better corrosion resistance, mechanical properties, and cytocompatibility than Ti-6Al-4V alloy[14]. Okazaki *et al.* showed that living bone came close to the Ti-15Zr-4Ta-4Nb alloy surface[14] but that the alloy showed marginal osteoconductivity due to low biological affinity; thus the alloy is regarded as bioinert. If the bioactivity of the Ti-15Zr-4Ta-4Nb surface is improved, this alloy is expected to be useful as a next generation bone substitute.

MATERIALS AND METHODS
Preparation of Specimens

Ti-15Zr-4Ta-4Nb alloy (Zr 15.36, Ta 3.95, Nb 3.95, Pb 0.18, Fe 0.26, O 0.23, N 0.04, Ti res. %) was prepared through a vacuum arc-melting process according to previous reports[14, 15]. After α–β forging, the alloy was annealed for 2 h at 700 °C. The surface and metallographic structure of the Ti-15Zr-4Ta-4Nb after thermal oxidation was studied on rectangular specimens of 10 x 10 x 3 mm[3] in size, after they were polished with emery paper #2000 (SiC grains, 0126 ± 0.02 μm Ra, SURFTEST SV-3000, Mitsutoyo Corporation, Tokyo, Japan). Each of them was then washed in an ultrasonic cleaner in ultra-pure water for 30 min and for another 30 min in acetone. Macro-grooves of 100~500 μm in both depth and width were machined on a surface of specimens 12 x 12 x 5 mm[3] in size in a manner similar to that described elsewhere[13]. These specimens were thermally oxidized at 500 °C for 1 h in air in a muffle furnace, and then cooled down to room temperature outside of the furnace in air. Fig. 1 shows the Ti-15Zr-4Ta-4Nb

1 cm

Figure 1 Ti-15Zr-4Ta-4Nb alloy treated with GRAPE® Technology

Placed upside-down in Kokubo's SBF

Micro-groove

Figure 2 The sample with micro-grooves, soaked in SBF.

alloy treated with GRAPE® Technology.

In Vitro Apatite-Forming Ability

Thermally oxidized specimens with micro-grooves were soaked in 30 cm^3 of SBF[16, 17] for 7 days kept at pH 7.4 and 36.5 °C. Fig. 2 shows how the sample was soaked: The machined surface was so placed as to face the bottom of flat-bottomed polystyrene bottles. By doing so, the specific narrow spaces designed above were achieved. The specimens were removed from SBF, then, gently rinsed with ultra-pure water, and dried at room temperature. The surfaces in the macro-grooves were observed by a field emission scanning electron microscope (FE-SEM, S-4700, Hitachi Co., Ltd., Tokyo, Japan).

In Vivo Tests

The specimens shown in Fig. 3(a) were used for *in vivo* tests. The specimens were implanted with strict aseptic surgical techniques and ordinary sterile preparation (autoclaving sterilization at 121 °C for 20 min, 0.11 Map). The procedure was strictly in accordance with the guidelines of the Animal Committee of the Graduate School of Medicine and Dentistry, Okayama University. A transcortical implant model was applied to Japanese white male rabbits (weight 2.6–3.0 kg). Premedication was given by intramuscular injection of ketamine hydrochloride (Ketalar) and xylazine hydrochloride (Selactar) into the gluteus maximus muscle. A straight incision was made in the area of the rabbit's tibia. Superficial fascia and periostreal membrane were incised using a sharp dissection to expose the bone surface. A drill 6.0 mm in diameter was used to make a hole (6.1 mm hole) in both tibiae of the rabbit at 10 mm distal from the knee joint. The specimen was seated using finger pressure and gently tapped into the space of the hole, followed by closing the skin with a nylon suture. A specimen treated by GRAPE® Technology was inserted in the hole of the right tibia, and a specimen with only grooves in the hole of the left tibia. After 3 weeks, the rabbits were sacrificed by inhalation of isoflurane and an excess intravenous injection of pentobarbital sodium solution. The tibiae with the inserted specimens were taken out and soaked in 10 % neutral buffered formaldehyde aqueous solution up to the next treatment. The specimens were subsequently dehydrated in an ascending series of ethyl alcohols and infiltrated with methylmethacrylate. The hardened blocks were positioned in a microtome and sectioned along the long axis of each sample to obtain sections of about 30 μm, shown in Fig. 3(b), then stained using toluidine blue for light microscopy analysis.

RESULTS

In Vitro Apatite-Forming Ability

Fig. 4 shows FE-SEM images of the internal surface (wall) of the micro-grooves after

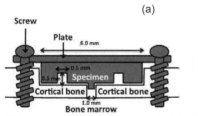 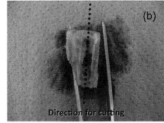

Figure 3 (a) The specimen in the transcortical implant model and (b) a sample retrieved after 3 weeks of implantation.

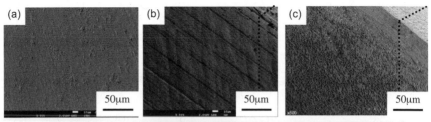

Figure 4   FE-SEM images on the internal surfaces of specimens with (a) only heating, (b) grooves (no heating), and (c) treated with GRAPE® Technology (micro-grooves and thermal oxidation at 500 °C for 1 h) after soaking in SBF for 7 d.

soaking in SBF for 7 days: (a) specimen with heating (no grooves), (b) with grooves (no heating), and (c) treated with GRAPE® Technology (grooves and thermal oxidation at 500 °C).   Hemispherical particles were observed for the specimen treated with GRAPE® Technology, but not for the other ones even with soaking longer than 7 days.   The micro-beam X-ray diffraction profile (Cu Kα) in Fig. 5 indicated the presence of 26° and 32° peaks, assignable to hydroxyapatite (JCPDS#09-0432), for the deposits found in Fig. 4(c).

In vivo Osteoconductivity

Apatite deposition in SBF, confirmed above, strongly suggests that the specimens treated with GRAPE® Technology should exhibit bone-cell ingrowth and bond to living bone tissues *in vivo*.   Fig. 6 shows the histological representation of the sections for micro-grooved Ti-15Zr-4Ta-4Nb alloys with and without thermal oxidation.   Light microscopic analysis of the histological sections demonstrated that new bone tissue formed in the micro-groove area after implantation for 3 weeks.   Fig. 6 (a) indicates a large amount of newly formed bone migrated into the groove space and is attached to the groove surface, *i.e.*, osteoconduction is realized.   In contrast, such bone ingrowth is hardly seen in Fig. 6 (b) where no cell-wall attachment is observed, and the groove space is even filled with soft tissue (light brownish area).   That is, without thermal oxidation, the bone tissue did not achieve direct contact with the specimen surface and hardly infiltrated into the grooves.

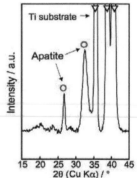

Figure 5   A micro-beam X-ray diffraction profile for the deposits in the groove Fig. 4(c).

DISCUSSION

Sugino *et al.*[17] confirmed the presence of rutile layer on Ti-15Zr-4Ta-4Nb thermally oxidized at 600 °C, but not on the specimen oxidized at 500 °C for which they confirmed the presence of an oxide layer.   Nevertheless, both of the 500° and 600 °C samples deposited apatite in their micro-grooves when soaked in SBF for 7 days.   Therefore, it is the presence of appropriately designed narrow spaces or spatial gaps that control the apatite formation or *in vivo* activity, but the crystallinity of the titanium oxide layer has little to do with such ability.

The histological results fully confirmed not only the *in vitro* apatite formation on the Ti-15Zr-4Ta-4Nb alloy treated with GRAPE® Technology but also the *in vivo* osteoconductivity.   It is commonly accepted that tissue-implant bonding is initiated by the formation of a bone-like apatite layer on the surface of implant materials[17].   Fig. 4 showed that GRAPE® Technology is applicable to inducing *in vitro* apatite deposition on the surface of the Ti-15Zr-4Ta-4Nb alloy when the micro-gaps

(a) 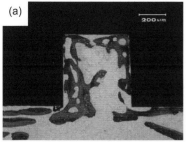 (b)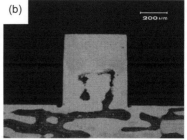

Figure 6   Histological sections after 3 weeks implantation.
(a) specimen treated with GRAPE® Technology   (b) specimen with only micro-grooves

and thermal treatment are properly designed, and to inducing osteoconductivity that leads to direct bonding to bony tissues.   GRAPE® Technology is probably the simplest method that has potential to improve the fixation of artificial joint materials to bone.   It is one of the greatest advantages for the technology to provide the surface with intrinsic osteoconduction ability, and thus, even if the apatite layer formed should be peeled off under certain conditions, the surface will be self-healed to spontaneously deposit apatite again.   The remarkable migration of newly formed bone into the grooves also suggests that GRAPE® Technology could improve the fixation strength between bone and the implanted alloy.

CONCLUSION

Ti-15Zr-4Ta-4Nb treated with GRAPE® Technology, for which micro-grooves of appropriate sizes, e.g., 500 μm in depth and width, were provided and thermally oxidized at 500 °C and 600 °C, induced the formation of apatite on the internal surface of the micro-grooves after soaking in SBF.   *In vivo* experiments confirmed the osteoconductivity of Ti-15Zr-4Ta-4Nb treated with GRAPE® Technology.

Acknowledgements

This study is partially supported by "Super Special Consortia" for the supporting development of cutting edge medical care (Cabinet Office, Government of Japan / Ministry of Health, Labour and Welfare / Ministry of Education, Culture, Sports, Science and Technology / Ministry of Economy, Trade and Industry.

REFERENCES
[1]C. Ohtsuki, H. Iida, S. Hayakawa, A. Osaka, Bioactivity of titanium treated with hydrogen peroxide solutions containing metal chlorides, *J. Biomed. Mater. Res.,* **35**, 39-47 (1997).
[2]S. Kaneko, K. Tsuru, S. Hayakawa, C. Ohtsuki, T. Ozaki, H. Inoue, A. Osaka, In vivo evaluation of bone-bonding of titanium metal chemically treated with a hydrogen peroxide solution containing tantalum chrolide, *Biomaterials,* **22**, 875-881 (2001).
[3]W. Jin-Ming, F. Xiao, S. Hayakawa, K. Tsuru, S. Takemoto, A. Osaka, Bioactivity of metallic biomaterials with anatase layers deposited in acidic titanium tetrafluoride solution, *J. Mater. Sci.: Mater. Med.,* **14**, 1-5 (2003).
[4]X.X. Wang, S. Hayakawa, K. Tsuru, A. Osaka, Improvement of bioactivity of H_2O_2/$TaCl_5$-treated titanium after subsequent heat treatments, *J. Biomed. Mater. Res.,* **52**, 171-176 (2000).

[5]X.X. Wang, S. Hayakawa, K. Tsuru, A. Osaka, A comparative study of in vitro apatite deposition on heat-, H_2O_2, and NaOH-treated titanium surfaces, *J. Biomed. Mater. Res.,* **52**, 172-178 (2001).

[6]X.X. Wang, W. Yan, S. Hayakawa, K. Tsuru, A. Osaka, Apatite deposition on thermally and anodically oxidized titanium surfaces in a simulated body fluid, *Biomaterials,* **24**, 4631-4637 (2003).

[7]A. Osaka, X.X. Wang, S. Hayakawa, K. Tsuru, Biomimetic deposition of apatite on electrochemically oxidized titanium substrates, *in Bioceramics, ed. S. Giannini and A. Moroni, Trans Tech Publications,* **Vol. 13,** 263-266 (2000).

[8]A. Osaka, S. Hayakawa, K. Tsuru, S. Takemoto, Y. Kawabe, S. Iwatani, In vitro biomimetic deposition of apatite on chemically and electrochemically treated titanium, *J. Aust. Ceram. Soc.,* **37(1),** 1-8 (2001).

[9]X. Liu, C. Ding, Z. Wang, Apatite formed on the surface of plasma-sprayed wollastonite coating immersed in simulated body fluid, *Biomaterials,* **22**, 2007-2012 (2001).

[10]J. Ma, C.H. Liang, L.B. Kong, C. Wang, Colloidal characterization and electrophoretic deposition of hydroxyapatite on titanium substrate, *J. Mater. Sci.: Mater. Med.,* **14(9)**, 797-801 (2003).

[11]T. Kokubo, H. Takadama, How useful is SBF in predicting in vivo bone bioactivity?, *Biomaterials,* **27**, 2907-2915 (2006).

[12]A. Sugino, K. Uetsuki, K. Tsuru, S. Hayakawa, C. Ohtsuki, A. Osaka, Gap effect on the heterogeneous nucleation of apatite on thermally oxidized titanium substrate, *Key. Eng. Mater.,* **361-363,** 621-624 (2008).

[13]A. Sugino, K. Uetsuki, K. Tsuru, S. Hayakawa, A. Osaka, C. Ohtsuki, Surface topography designed to provide osteoconductivity to titanium after thermal oxidation, *Mater. Trans.,* **49**, 428-434 (2008).

[14]Y. Okazaki, S. Rao, Y. Ito, T. Tateishi, Corrosion resistance, mechanical properties, corrosion fatigue strength and cytocompatibility of new Ti alloys without Al and V, *Biomaterials,* **19**, 1197-1215 (1998).

[15]Y. Okazaki, E. Nishimura, H. Nakada, K. Kobayashi, Surface analysis of Ti-15Zr-4Ta alloy after implantation in rat tibia, *Biomaterials,* **22**, 599-607 (2001).

[16]T. Kokubo, H. Kushitani, S. Sakka, T. Kitsugi, T. Yamamoto, Solution able to reproduce in vivo surface-structure changes in bioactive glass-ceramics A-W^3, *J. Biomed, Mater. Res.,* **24**, 721-734 (1990).

[17]A. Sugino, C. Ohtsuki, K. Tsuru, S. Hayakawa, T. Nakano, Y. Okazaki, A. Osaka, Effect of spatial design and thermal oxidation on apatite formation on Ti-15Zr-4Ta-4Nb alloy, *Acta. Biomater.,* **5**, 298-304 (2009).

RAPID BIOMIMETIC CALCIUM PHOSPHATE COATING ON METALS, BIOCERAMICS AND BIOPOLYMERS AT ROOM TEMPERATURE WITH 10xSBF

A. Cuneyt Tas
Department of Biomedical Engineering, Yeditepe University
Istanbul 34755, Turkey

ABSTRACT
This paper reports the utilization of high ionic strength (>1100 mM) calcium phosphate solutions in depositing 20-65 μm-thick, bonelike apatitic calcium phosphate on Ti6Al4V within 2 to 6 hours, at room temperature. The solution used here multiplied the concentrations of calcium and phosphate ions in human plasma or synthetic/simulated body fluid (SBF) by a factor of ten. The solutions did not contain any buffering agents, such as Tris or Hepes. With these solutions there was no CO_2 bubbling required. The carbonate content (8 wt%) and Ca/P molar ratio (1.57) of the coated calcium phosphates qualified them as bonelike. The same solutions were successfully used in coating macroporous β-TCP (β-tricalcium phosphate, $β-Ca_3(PO_4)_2$) cylinders and macroporous collagen membranes with apatitic calcium phosphate.

INTRODUCTION
SBF (synthetic/simulated body fluid) solutions are shown [1-3] to induce apatitic calcium phosphate formation on metals, ceramics or polymers (with proper surface treatments) soaked in them. SBF solutions, in close resemblance to the Hanks Balanced Salt Solution (HBSS) [4], are prepared with the aim of simulating the ion concentrations present in the human plasma. It is noted that physiological HBSS solutions are also able to induce apatite formation on titanium [5]. To mimic human plasma, SBF solutions are prepared to have relatively low calcium and phosphate ion concentrations, namely, 2.5 mM and 1.0 mM, respectively [6]. Furthermore, to mimic human plasma, pH value of SBF solutions was adjusted to the physiological value of 7.4 by using organic buffers, such as Tris [3] or Hepes [7]. These compounds are not present in human plasma. The buffering agent Tris present in conventional SBF solutions, for instance, is reported [8] to form soluble complexes with several cations, including Ca^{2+}, which further reduces the concentration of free Ca^{2+} ions available for coating. Hepes, on the other hand, is rather unstable and easily loses a certain fraction of carbonate ions [9]. The hydrogencarbonate ion (HCO_3^-) concentration in SBF solutions was kept between 4.2 mM (equal to that of HBSS) and 27 mM [7, 9, 10].
Having their ionic compositions more or less similar to that of human blood plasma, HBSS or SBF formulations have only a limited power with respect to the precipitation of apatitic calcium phosphates. As a direct consequence, nucleation and precipitation of calcium phosphates from HBSS or SBF solutions are rather slow [11]. To get total surface coverage of a 10 x 10 x 1 mm titanium or titanium alloy substrate immersed into a 1.5 or 2 x SBF solution, one typically needs to wait for 2 to 3 weeks, with frequent (at every 36 to 48 hours) replenishment of the solution [12]. The broad motivation in this work is to enhance the kinetics of coating deposition.
In order to achieve the above objective, Barrere et al. [13-17] have developed 5xSBF-like solution recipes (with pH values close to 5.8), which did not employ any buffering agent, such as Tris or Hepes. In these studies [13-17], coating was achieved by employing two different solutions (solutions A and B as they referred), and pH was adjusted by continuous bubbling of CO_2 gas into the reaction chamber. A coating thickness of about 30 μm was achieved only after 6 h of immersion, which did not increase much even after 48 hours of further soaking, stirring and constant CO_2 bubbling at 50°C [13]. Moreover, they also introduced additional intermediate steps. These included [13] immersing the metal strips in the first 5xSBF solution (to seed the surface with calcium phosphate

nuclei) for 24 h at 37°C, followed by another soaking in their second 5xSBF solution (to form the actual coat layers by a so-called growth process) for 6 to 48 h at 50°C [13]. Dorozhkin *et al.* [18] modified this CO_2-bubbling technique, by using two different "4xSBF" solutions instead. These additional intermediate steps and second solution treatments add extra time and oppose the advantage gained by the enhanced kinetics.

There is yet another concern over the above-mentioned CO_2-bubbling technique. Bubbling of CO_2 (with the sole purpose of maintaining the solution pH at around neutral values, through concentrated SBF-like solutions) results in calcium phosphate coatings with significantly increased carbonate ion concentrations. A quantitative evidence for this phenomenon was provided by Dorozhkina *et al.* [11]. Dorozhkina *et al.* [11] reported the CO_3^{2-} weight percentages of 19, 26 and 33 (in the resultant calcium phosphates) for 2, 4 and 8xSBF solutions, respectively, when CO_2 bubbling was used for pH regulation at 37°C. They also noted that the same samples with such high carbonate concentrations also exhibited Ca/P molar ratios of 1.8, 1.9 and 2.3, respectively. For comparison's sake, human bones contain 7.4% CO_3^{2-}, 34.8% Ca^{2+}, 46.6% PO_4^{3-}, 0.72% Mg^{2+} and around 10.5% H_2O, by weight [11]. The Ca/P molar ratio of human bones is around 1.75 [19, 20]. Therefore, a calcium phosphate-based material with a Ca/P molar ratio of 2.3 and a carbonate content of about 33 wt% can not be regarded as a "bonelike" substance.

The aim of this paper is to present the preparation of a new acidic solution, which contains 10 times the calcium and phosphate ion concentrations of human blood plasma. Such a solution should enhance the kinetics of coating formation even more. Furthermore, it is preferred that other than the surface treatment step, not too many intermediate steps are involved. The only step that is needed is to add $NaHCO_3$ into the solution to raise its pH to around 6.5. The resultant solution is able to coat Ti6Al4V strips, macroporous β-TCP cylinders and porous collagen membranes (at RT, 22±1°C) rapidly, in as little as 2 hours. It was hereby shown that it is not necessary to use those so-called biomimetic conditions (i.e., 37°C and pH 7.4) for the coating purposes.

EXPERIMENTAL PROCEDURE
Preparation of Ti6Al4V Strips
    Sheets of Ti6Al4V (Grade 5 ASTM B265) were cut into rectangular strips with typical dimensions of 10 x 10 x 0.20 mm and first abraded manually with a 1200-grit SiC paper. Strips were then cleaned with acetone (15 min), ethanol (15 min) and deionized water (rinsing), followed by etching each strip in 150 mL of a 5 M KOH solution at 60°C for 24 h, in a sealed glass bottle. Thoroughly rinsed (w/deionized water) strips were finally heat-treated at 600°C for 1 h in clean Al_2O_3 boats, with heating and cooling rates of 3°C/min.
Preparation of Macroporous β-TCP Cylinders
    Macroporous, trabecular hydroxyapatite (of bovine origin, Endobon®, Merck Biomaterials GmbH, Darmstadt, Germany) cylinders were used to form macroporous β-TCP samples. One such cylinder (approx. 3 cm in height and 1 cm diameter, weighing 3.5 g) was placed into a 100 mL-capacity glass media bottle containing 90 mL deionized water and 18.0 g $(NH_4)_2HPO_4$, and the bottle was sealed prior to placing it into a 60°C oven. The cylinders were removed from the media bottles after 4 h at 60°C, and kept in a drying oven at 60°C (the cylinder was not washed with water) in unused Al_2O_3 dishes for 12 h. The dried samples (in the same Al_2O_3 dishes) were finally heated to 1225°±3°C at the heating rate of 5°C/min, soaked at this temperature for 12 h and cooled back to the room temperature at the rate of 2°C/min to obtain β-TCP cylinders.
Porous Collagen Membranes
    Approximately three millimeter-thick collagen membranes or sponges (i.e., Matrix Collagen Sponge™) were obtained from Collagen Matrix, Inc. (Franklin Lakes, NJ, USA), and used as-received after cutting those into 1 x 1 x 0.3 cm coupons. These sponges were consisting of purified Type I

collagen (which in turn is rich in glycine, proline, and hydroxyproline) in its native triple helical structure.

Coating solutions

Solution preparation recipe (for a total aqueous volume of 2 L) is given in Table 1. The chemicals given in Table 1 are added, in the order written, to 1900 mL of deionized water in a glass beaker of 3.5 L-capacity. Before the addition of the next chemical, the previous one was completely dissolved in deionized water. After all the reagents were dissolved at RT, the solution volume was completed to 2 L by adding proper amount of deionized water. This extremely stable (against the formation of any visible calcium phosphate precipitates) stock solution of pH value of 4.35-4.40 can be stored at RT, in a capped glass bottle, for more than a year without precipitation.

Coating of Samples

Just prior to coating a Ti6Al4V strip, a 200 mL portion of the above-mentioned stock solution was placed into a 250 mL-capacity glass beaker, and a proper amount of $NaHCO_3$ powder was added to raise the hydrogencarbonate ion (HCO_3^-) concentration to 10 mM, under vigorous stirring. Following the rapid dissolution of the $NaHCO_3$, the pH of the clear solution automatically rose to 6.50 at RT. This solution (with an ionic strength of 1137.5 mM) was then transferred to a 250 mL-capacity glass bottle, which contained the Ti6Al4V strip inside, tightly capped and kept at RT for 2 to 6 hours during *in situ* coating. For coating the macroporous β-TCP cylinders and the porous collagen membranes, the same procedure was repeated in 250 mL-capacity glass media bottles strictly in the same manner as described for the Ti6Al4V strips.

Table I   Stock solution preparation recipe, for a total volume of 2 L

Reagent	Order	Amount (g)	Concentration (mM)
NaCl	1	116.8860	1000
KCl	2	0.7456	5
$CaCl_2·2H_2O$	3	7.3508	25
$MgCl_2·6H_2O$	4	2.0330	5
NaH_2PO_4	5	2.3996	10

Sample Characterization

After the experiments were over, the samples were taken out of the solutions and rinsed respectively with an ample supply of deionized water and 80% ethanol solutions, followed by drying in air in a well-circulated fume hood for 24 h. Samples were characterized by XRD (D8 Advance, Bruker AXS, Karlsruhe, Germany; operated with Cu K_α radiation at 40 kV and 40 mA), FTIR (SpectrumOne, Perkin-Elmer, MA), SEM-EDXS (Hitachi S-4700 in the secondary electron mode, acceleration voltage 5-15 kV), and ICP-AES (Thermo Jarrell Ash, Model 61E, Woburn, MA). Gold sputtering was employed to make the coating surfaces conductive for the SEM investigations. In order to measure the thickness of the coat layers, the metallic strips were tilted by 45 degrees and studied by SEM.

RESULTS AND DISCUSSION

The chemical and thermal treatment of Ti6Al4V strips prior to the coating runs were mainly performed according to the previously published methods [6, 23, 24]. However, in our modification to the alkali treatment procedure, we have used 5 M KOH solution instead of 5 M NaOH. Figure 1a showed the surface of 5 M KOH + 600°C treated Ti6Al4V. The aggregated rosettes seen on the surface (Fig. 1a) belong to a potassium titanate phase of a possible composition of $K_2Ti_5O_{11}$. It should be pointed out that this tentative formula is only based on the quantitative SEM-EDXS analyses performed on the rosettes seen in Fig. 1a. A phase of similar stoichiometry (i.e., $Na_2Ti_5O_{11}$) was also observed in case of using 5 M NaOH+600°C-treatment [24]. The surface of the alkali- and heat-treated strips also contained rutile (TiO_2), as seen in the XRD chart of Fig. 1b. The peak positions (labeled with "1") for the potassium titanate phase of the XRD chart in Fig. 1b match well with those reported previously, for sodium titanate, by Kim *et al.* [24]. Masaki *et al.* [25] recently reported the complete conversion of Ti metal powders into $KTiO_2(OH)$, upon soaking the metal powders in a concentrated (>35 M), hot (150°C) bath of KOH. Masaki *et al.* [25] also noted that this new phase transformed at 528°C into $K_2Ti_2O_5$, when heated in air. On the other hand, Yuan *et al.* [26] reported that TiO_2 powders heated in an 8 M KOH solution first formed $K_2Ti_{18}O_{17}$ nanowires, which would then decompose into $K_2Ti_6O_{13}$ and TiO_2 upon calcination in air at 600°C.

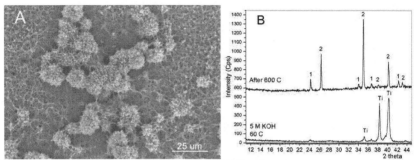

Figure 1 (a) Surface of 5 M KOH + 600°C-treated Ti6Al4V strips prior to coating, (b) XRD data of 5M KOH- (*trace-a*) and 5M KOH+600°C-treated Ti6Al4V strips; *phase 1: potassium titanate, phase 2: rutile, TiO_2.*

K+ ions originating from the potassium titanates formed on the surfaces of KOH- and heat-treated Ti6Al4V strips, when exposed to the coating solution, were released into the solution in exchange of H_3O^+ ions, and eventually resulting in the formation of a Ti-OH layer. Ca^{2+} ions from the coating solution were then incorporated in this basic layer and act as embryonic sites for the nucleation of carbonated apatitic calcium phosphates [23]. The coating solution described above was not stable against precipitation (at RT) after the addition of $NaHCO_3$ to raise its pH to the vicinity of 6.5. The rise of pH in these solutions was quite monotonical (Fig. 2a). pH versus time curve depicted in Figure 2a was obtained after adding 1.68 g $NaHCO_3$ (i.e., 10 mM HCO_3^-), in powder form, to the solution (of pH 4.4 at 22°C) given in Table I. The stability against homogeneous precipitation only lasted from 5 to 10 minutes at RT, following the addition of $NaHCO_3$. After that period, solutions containing the metal strips slowly started displaying turbidity (from 10 minutes to the end of the first hour), and by the end of 2 hours the solution turned opaque. The colloidal calcium phosphate nuclei formed in the solution stay suspended, and could only be separated from the mother liquor by centrifugal filtration (>3000 rpm). However, it is interesting to note that the solution pH at the end of 2 hours of soaking period stayed the same or slightly increased to around 6.57 or 6.58. This slight increase in pH was ascribed to the release of CO_2 [14]. A pH decrease would have been encountered during the formation of colloidal precipitates due to H+ release, but such a pH drop is not always observed [14, 27]. To perform a run with 6 hours of total soaking time, the coating solution for the same strip was replenished twice with a new transparent solution (of pH = 6.5) at the end of each 2-hours segment (see the XRD data of Fig. 2b below). The start of precipitation indicated the stage where the solution reached supersaturation.

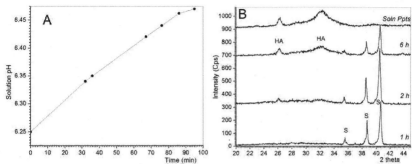

Figure 2   (a) pH *vs* time curve of solution given in Table I immediately after the addition of 1.68 g NaHCO$_3$ (*at 22 °C*), (b) XRD data of coated Ti6Al4V strips as a function of soaking time (S: peaks of substrate) and those of colloidal calcium phosphate precipitates

It must be noted that the extremely simple solution recipe and the robust coating process described in this study utilized a HCO$_3^-$ concentration of only 10 mM. This number is significantly lower than those used in "5xSBF + continuous CO$_2$-bubbling" method [13-18], mainly because in the latter, the coating needed to be continued for at least 72 h at 37° or 50°C, under continuous bubbling of CO$_2$ [13]. Due to this long time of deposition, attention must be paid there to ensure that coating deposition is linear. Increased carbonate concentration in a coating solution would result in calcium phosphate-like solid deposits with unacceptably high levels (25 to 30 wt%) of carbonate ions [11].

The coating process reported in this paper, on the contrary, does not require any attention. The inexpensive and stable solution given in Table I is simply prepared, then 10 mM NaHCO$_3$ is added to it at once in powder form, the solution and sample to be coated is placed in a sealed, undisturbed glass container, and after keeping it at RT for 2 to 6 hours, the sample is *in situ* coated by an apatitic calcium phosphate layer. Table II shows the deposition rate (measured in terms of coating thickness) as a function of immersion/soaking time.

Table II   Coating thickness as a function of soaking time at RT in 10xSBF

Soak time (*h*)	Coating thickness (*μm*)
1	13 ± 2
2	22 ± 2
4	46 ± 4
6	68 ± 5

Such a linear and enhanced coating rate has never been achieved before by using either 1.5xSBF or 5xSBF solutions. With the use of 5xSBF solutions under constant CO$_2$ bubbling, the maximum coating thickness attained was around 35 μm after 3 days (i.e., 72 hours) of deposition [13-17].

Figure 3 depicts the SEM photomicrographs of coated surfaces of Ti6Al4V strips as a function of coating time (1 to 6 h; Figs. 3a through 3d) at RT. By using 10xSBF solutions described here, the surface of the KOH- and 600°C-treated surface of Ti6Al4V strips are rapidly covered within the first hour of immersion (Fig. 3a) with a smooth, nano-textured calcium phosphate layer of about 13 μm-thick. By the end of the second hour in solution, coating develops to a thickness of about 22 μm, however, the attachment of calcium phosphate globules onto that initially-formed smooth surface becomes more enhanced (Fig. 3b). Such globules of apatitic calcium phosphate were quite similar to the previously reported results relevant to biomimetic SBF coating, excepting that biomimetic conditions (i.e., pH 7.4 and 37°C) were not met in our study. SEM micrograph given in Figure 3f is supplied for comparison purposes only. It was recorded from a Ti6Al4V strip soaked in *Tas*-SBF (a Tris-buffered SBF of pH 7.4, with a HCO_3^- concentration equal to 27 mM [10]) for 2 weeks at 37°C. A conventional SBF solution (i.e., 1.5x*Tas*-SBF) can only coat a 20 μm-thick layer of apatitic calcium phosphate after two weeks of soaking at 37°C, while the 10 SBF reported here achieves this in only 2 hours at RT. High-magnification photomicrographs of Figures 3c and 3d showed that the globules actually consisted of petal-like nanoclusters.

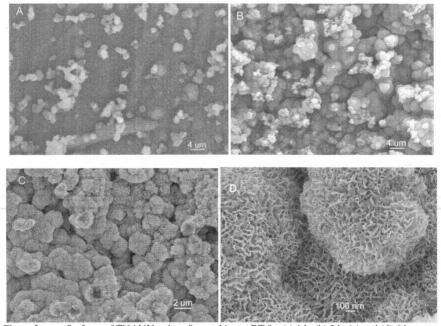

Figure 3   Surfaces of Ti6Al4V strips after soaking at RT for (a) 1 h, (b) 2 h, (c) and (d) 6 h

The FTIR data of the solution precipitates (chart not shown) indicated the nano-crystalline apatitic calcium phosphates formed in the solution, with characteristic IR bands of the O–H stretching and bending of H_2O at 3440 and 1649 cm⁻¹. Presence of carbonate groups was confirmed by the bands at 1490-1420 and 875 cm⁻¹. PO_4 bands were recorded at 570 and 603 (v_4), 962 (v_1), 1045 and 1096 (v_3) cm⁻¹ [10]. It is important to note that neither the precipitates themselves nor the coating layer (on Ti6Al4V strips) contained $CaCO_3$ (calcite) [28].

If the sole aim of a process is to coat titanium or titanium alloy surfaces with a carbonated apatitic calcium phosphate layer, then there is no need to maintain the pH value of a coating solution exactly at the physiologic value of 7.4. This point has been successfully confirmed in the work of

Barrere *et al.* [13-17, 29]. One only needs to be aware of the delicate balance between the solution pH, HCO_3^- ion concentration and temperature in determining which phases will be soluble or not under a specific set of those conditions [30]. On the other hand, the presence of TRIS or HEPES (added for the sole purpose of fixing the solution pH at around 7.4) in an SBF formulation simply retards the coating process to the level that in order to obtain a decent surface coverage one needs to wait for 1 or 3 weeks [6-8, 24].

Fast coating solutions, sometimes named as supersaturated calcification solutions (SCS) are not new either; for instance, the pioneering work of Wen *et al.* [31] showed that even in a TRIS-buffered SCS solution it would be possible to form 16 µm-thick calcium phosphate coat layers in after 16 hours of immersion. Choi *et al.* [32] reported the room temperature coating (about 10 µm-thick in 24 hours) of nickel-titanium alloy surfaces by a simple SCS solution, which was not even buffered at the physiologic pH. The present paper corroborates these previous findings and reports further improvements.

It is known that an amorphous calcium phosphate (ACP) precursor is always present during the precipitation of apatitic calcium phosphates from the highly supersaturated solutions, such as the one used here [33]. Posner, *et al.* [34] proposed that the process of ACP formation in solution involved the formation first of $Ca_9(PO_4)_6$ clusters which then aggregated randomly to produce the larger spherical particles or globules (as seen in Figs. 2d and 4), with the intercluster space filled with water. Such clusters (with a diameter of about 9.5 Angstrom [33]), we believe, are the transient solution precursors to the formation of carbonated globules with the stoichiometry of a calcium-deficient hydroxyapatite, namely, $Ca_{10-x}(HPO_4)_x(PO_4)_{6-x}(OH)_{2-x}$, where x might be converging to 1 [14].

Ca/P molar ratio of the coat layers (after scraping small portions of the coatings off of the Ti6Al4V strips) was measured by ICP-AES analysis. The samples collected were carefully ground into a fine powder, followed by dissolving those in a concentrated acid solution prior to the ICP-AES runs. Ca/P molar ratio in these samples turned out to be 1.57 ± 0.05. Carbonate content was found to be less than 10 wt% (i. e., $8.2 \pm 0.3\%$). This means that the deposited material consists of "carbonated, calcium-deficient, poorly crystallized hydroxyapatite." This is how DeGroot and Kokubo [35] defined, back in 1994, the material coated on a titanium substrate immersed in a conventional SBF solution as "bonelike." From this viewpoint, the present coatings can be classified as bonelike.

Onuma *et al.* [36] have demonstrated, by using dynamic light scattering, the presence of calcium phosphate clusters from 0.7 to 1.0 nm in size in clear simulated body fluids. They reported that calcium phosphate clusters were present in SBF even when there was no precipitation. This was true after 5 months of storage at RT. The solution coating procedure described here probably triggered the hexagonal packing [36] of those nanoclusters to form apatitic calcium phosphates, just within the first 5 to 10 minutes, following the introduction of $NaHCO_3$ to an otherwise acidic calcium phosphate solution. Since these nanoclusters are always present even in a conventional ionic strength SBF, the insertion of a suitable alkali- and heat-treated Ti6Al4V surface into such a solution immediately starts the coating process, as explained above. This is how the dense-looking under coat layer is formed (Figs. 3a, 3b, and 3f) in less than an hour, provided that the solution is concentrated and supplying enough Ca and HPO_4 ions to the metal-solution interface. What is achieved here, with this new solution in less than an hour, can only be done with a conventional SBF in about a week. On the other hand, the colloidal precipitates (as a result of the hexagonal packing of the invisible nanoclusters [36]) of 10 SBF solution are formed by a homogeneous nucleation process. The presence of these precipitates within the solution, possibly, further accelerates the coarsening of the newly deposited calcium phosphate globules. Conventional, Tris or Hepes-buffered SBF solutions (1.5 SBF) are able to form those precipitates by the end of 2nd or 3rd day of soaking at 37°C. Since the Ca/P molar ratio of all SBF solutions (including the one presented here) are 2.50, they are not stable against hydroxyapatite precipitation when the solution pH is higher than 6.4 [10, 11].

Figure 4 showed the XRD data of macroporous bioceramic cylinders (initially HA, finally β-TCP with a low HA content), as a function of increasing calcination time from 900° to 1225°C. The samples shown in this figure were prepared in the manner described in the Experimental chapter, the only difference among them is the temperature of calcination. Figure 5a depicted the macroporous nature of the $(NH_4)_2HPO_4$-treated and 1225°C-heated bovine cylinders. The procedure used here to transform the HA-based macroporous bovine cylinders to β-TCP was simply adapted from the procedure described by Bauer [37]. It is important to note that the 10xSBF solution of this study was able to significantly cover the available bioceramic surfaces in a time as short as 6 h.

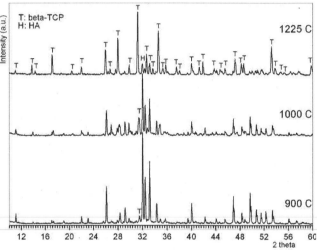

Figure 4   XRD traces of bovine-origin cylinders first $(NH_4)_2HPO_4$-treated and then heated at the temperatures shown

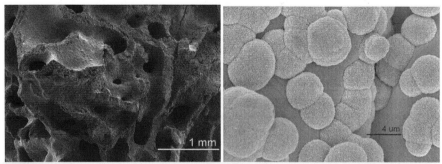

Figure 5   Surfaces of bovine bioceramics (a) after $(NH_4)_2HPO_4$ treatment, (b) 10xSBF-coated in 6h

ICP-AES analysis performed on six of the TCP-transformed cylinders, after 1225°C-heating, revealed that their Ca/P molar ratio was 1.52±0.015. The same ratio for the starting Endobon® cylinders was found to be in the vicinity of 1.70. The nanoporous CaP spherulites shown in Fig. 5 would also serve to increase the BET surface area of such macroporous β-TCP samples. In other words, having a β-TCP implant with an increased surface area will always be better, in terms of its osteointegrative properties, than having a β-TCP implant with low surface area. The follow-up studies will experimentally verify this assertion.

The characteristic SEM morphology of the as-received collagen membranes used in this study was shown in Fig. 6a. The typical microstructure after coating the membrane with apatitic calcium phosphate, by soaking in 10xSBF for 6 h, was given in Fig. 6b.

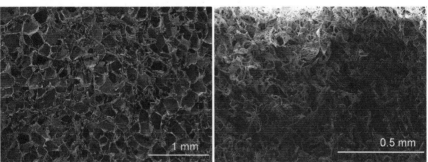

Figure 6   Collagen membrane microstructure (a) as-received, (b) 10xSBF soaked, 6 h at RT

The SEM photomicrographs of Figure 7 below depicted the microstructure of the 10xSBF-soaked collagen membranes at magnifications higher than those given in Figure 6. The procedure described here is one of the easiest ways of producing collagen-apatitic calcium phosphate composites. The membranes preserved their original flexibility after the 10xSBF coating. The rapid coating process also eliminated the problem of collagen swelling.

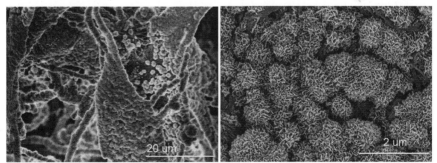

Figure 7   Collagen membranes soaked in 10xSBF solutions for 6 h at RT (*higher magnifications*)

Figure 8 showed the XRD data collected directly from the collagen membranes soaked in 10xSBF solution for 6 h at RT. This XRD chart was characteristic for the nanocrystalline (or poorly crystallized) nature of apatitic CaP formed in aqueous solutions of neutral pH. The data of Fig. 8 was collected from a 10xSBF-soaked membrane after placing it directly on the sample holder of the diffractometer.

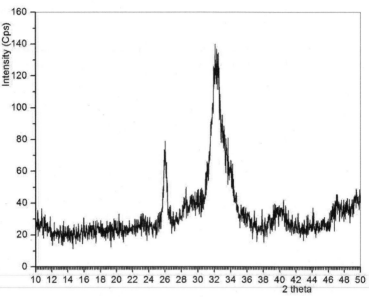

Figure 8   The typical XRD chart of 10xSBF-treated collagen membranes (for 6 h at RT)

Collagen is the *de facto* biopolymer in mammalians, whereas cellulose is that of plants and trees, and chitin is usually seen in the exoskeleton of crustaceans (sea-shells, oysters, crabs, etc.). Human bones are meticulous composites of collagen and carbonated, apatitic CaP mineralites. Collagen is tough and inextensible, with great tensile strength, and is the main component of cartilage, ligaments, tendons, and bones [38]. Multiple tropocollagen (with triple helix structures) molecules (300 nm long and 1.5 nm in diameter) form collagen fibrils, and multiple collagen fibrils form collagen fibers [39]. Soaking of collagen membranes or sponges in Tris–SBF-4.2 mM solution was first studied and reported by Rhee and Tanaka [40]. Lickorish *et al.* [41] later confirmed the findings of Rhee and Tanaka [40]. Girija *et al.* [42] tested the viability of a Hepes–NaOH-27 mM SBF solution [43] in synthesizing collagen–CaP composites by soaking fibrous collagen in that solution.

However, the process of 10xSBF-soaking of collagen membranes described here was unprecedented according to the best of our knowledge.

CONCLUSIONS

The use of $NaHCO_3$ with a concentrated (10 times of Ca^{2+} and HPO_4^{2-} ion concentrations) synthetic/simulated body fluid-like solution of ionic strength of 1137.5 mM allowed the formation of a bonelike apatitic calcium phosphate layer on Ti6Al4V at room temperature, within 2 to 6 hours. The coating solutions of pH 6.5 did not necessitate the use of buffering agents. The pH adjustment was achieved by a single-step addition of $NaHCO_3$. The coating process did not require the continuous bubbling of CO_2 during the process. This robust process had a linear and fast coating kinetics. The surfaces of the Ti6Al4V strips were chemically etched in 5 M KOH solution and thermally treated afterwards at 600°C, prior to soaking in 10xSBF solutions. KOH soaking and thermal treatment following it ensured the formation of potassium titanates on the strip surfaces. The coatings had a Ca/P molar ratio of 1.57 and contained 8 wt% CO_3^{2-}. Formation of colloidal nuclei, within the solution, was observed during the first hour of soaking at RT, but apparently the presence of those fine nuclei did not adversely affect the coating process.

Macroporous and bovine-origin HA cylinders were first converted into β-TCP, and then soaked in 10xSBF at RT for 6 h under conditions very similar to those employed in coating the Ti6Al4V strips. 10xSBF solutions were able to coat most of the available β-TCP surfaces.

Porous collagen (Type I) membranes were soaked at RT (for 6 h) in 10xSBF solutions and the collagene membranes were readily coated with apatitic and nanotextured calcium phosphate globules or spherulites. This process is a simple, rapid and robust way of producing porous collagen-apatitic calcium phosphate composites at room temperature within few hours.

REFERENCES

[1]T. Kokubo, Surface Chemistry of Bioactive Glass-Ceramics, *J. Non-Cryst. Solids*, **120**, 138-51 (1990).

[2]T. Kokubo, Apatite Formation on Surfaces of Ceramics, Metals and Polymers in Body Environment, *Acta Mater.*, **46**, 2519-27 (1998).

[3]T. Kokubo, H. M. Kim, M. Kawashita, and T. Nakamura, Bioactive Metals: Preparation and Properties, *J. Mater. Sci. Mater. Med.*, **15**, 99-107 (2004).

[4]J. H. Hanks and R. E. Wallace, Relation of Oxygen and Temperature in the Preservation of Tissues by Refrigeration, *Proc. Soc. Exp. Biol. Med.*, **71**, 196 (1949).

[5]L. Frauchiger, M. Taborelli, B. O. Aronsson, and P. Descouts, Ion Adsorption on Titanium Surfaces Exposed to a Physiological Solution, *Appl. Surf. Sci.*, **143**, 67-77 (1999).

[6]H. M. Kim, H. Takadama, F. Miyaji, T. Kokubo, S. Nishiguchi, and T. Nakamura, Formation of Bioactive Functionally Graded Structure on Ti-6Al-4V Alloy by Chemical Surface Treatment, *J. Mater. Sci. Mater. Med.*, **11**, 555-59 (2000).

[7]A. Oyane, K. Onuma, A. Ito, H. M. Kim, T. Kokubo, and T. Nakamura, Formation and Growth of Clusters in Conventional and New Kinds of Simulated Body Fluids, *J. Biomed. Mater. Res.*, **64A**, 339-48 (2003).

[8]A. P. Serro and B. Saramago, Influence of Sterilization on the Mineralization of Titanium Implants Induced by Incubation in Various Biological Model Fluids, *Biomaterials*, **24**, 4749-60 (2003).

[9]E. I. Dorozhkina and S. V. Dorozhkin, Surface Mineralisation of Hydroxyapatite in Modified Simulated Body Fluid (mSBF) with Higher Amounts of Hydrogencarbonate Ions, *Colloid. Surface. A*, **210**, 41-48 (2002).

[10]A. C. Tas, Synthesis of Biomimetic Ca-hydroxyapatite Powders at 37°C in Synthetic Body Fluids, *Biomaterials*, **21**, 1429-38 (2000).

[11]E. I. Dorozhkina and S. V. Dorozhkin, Structure and Properties of the Precipitates Formed from Condensed Solutions of the Revised Simulated Body Fluid, *J. Biomed. Mater. Res.*, **67A**, 578-81 (2003).

[12]H. Takadama, H. M. Kim, T. Kokubo, and T. Nakamura, TEM-EDX Study of Mechanism of Bonelike Apatite Formation on Bioactive Titanium Metal in Simulated Body Fluid, *J. Biomed. Mater. Res.*, **57**, 441-48 (2001).

[13]P. Habibovic, F. Barrere, C. A. van Blitterswijk, K. de Groot, and P. Layrolle, Biomimetic Hydroxyapatite Coating on Metal Implants, *J. Am. Ceram. Soc.*, **85**, 517-22 (2002).

[14]F. Barrere, C. A. van Blitterswijk, K. de Groot, and P. Layrolle, Influence of Ionic Strength and Carbonate on the Ca-P Coating Formation from SBF x 5 Solution, *Biomaterials*, **23**, 1921-30 (2002).

[15]F. Barrere, C. A. van Blitterswijk, K. de Groot, and P. Layrolle, Nucleation of Biomimetic Ca-P Coatings on Ti6Al4V from SBF x 5 Solution: Influence of Magnesium, *Biomaterials*, **23**, 2211-20 (2002).

[16]F. Barrere, C. M. van der Valk, R. A. J. Dalmeijer, C. A. van Blitterswijk, K. de Groot, and P. Layrolle, *In vitro* and *in vivo* Degradation of Biomimetic Octacalcium Phosphate and Carbonate Apatite Coatings on Titanium Implants, *J. Biomed. Mater. Res.*, **64A**, 378-87 (2003).

[17]F. Barrere, C. M. van der Valk, G. Meijer, R. A. J. Dalmeijer, K. de Groot, and P. Layrolle, Osteointegration of Biomimetic Apatite Coating Applied onto Dense and Porous Metal Implants in Femurs of Goats, *J. Biomed. Mater. Res.*, **67B**, 655-65 (2003).

[18]S. V. Dorozhkin, E. I. Dorozhkina, and M. Epple, A Model System to Provide Good *in vitro* Simulation of Biological Mineralization, *Cryst. Growth Design*, **4**, 389-95 (2004).

[19]M. C. Dalconi, C. Meneghini, S. Nuzzo, R. Wenk, and S. Mobilio, Structure of Bioapatite in Human Foetal Bones: An X-ray Diffraction Study, *Nucl. Instrum. Meth. B*, **200**, 406-10 (2003).

[20]R. Z. LeGeros, *Calcium Phosphates in Oral Biology and Medicine*, Karger Publications, Basel, Switzerland, 1991. pp. 4-44; 108-14.

[21]Designation: C-633. Standard Test Method for Adhesion Strength of Flame-sprayed Coatings, Annual Book of ASTM Standards, Vol. 3.01. Philadelphia, PA: American Society for Testing and Materials, pp. 665-69 (1993).

[22]T. Kokubo, F. Miyaji, H. M. Kim, and T. Nakamura, Spontaneous Formation of Bonelike Apatite Layer on Chemically Treated Titanium Metals, *J. Am. Ceram. Soc.*, **79**, 1127-29 (1996).

[23]L. Jonasova, F. A. Mueller, A. Helebrant, J. Strnad, and P. Greil, Biomimetic Apatite Formation on Chemically Treated Titanium, *Biomaterials*, **25**, 1187-94 (2004).

[24]H. M. Kim, F. Miyaji, T. Kokubo, and T. Nakamura, Effect of Heat Treatment on Apatite-forming Ability of Ti Metal Induced by Alkali Treatment, *J. Mater. Sci. Mater. Med.*, **8**, 341-47 (1997).

[25]N. Masaki, S. Uchida, H. Yamane, and T. Sato, Characterization of a New Potassium Titanate, $KTiO_2(OH)$ Synthesized via Hydrothermal Method, *Chem. Mater.*, **14**, 419-24 (2002).

[26]Z. Y. Yuan, X. B. Zhang, and B. L. Su, Moderate Hydrothermal Synthesis of Potassium Titanate Nanowires, *Appl. Phys. A*, **78**, 1063-66 (2004).

[27]P. A. A. P. Marques, M. C. F. Magalhaes, and R. N. Correia, Inorganic Plasma with Physiological CO_2/HCO_3^- Buffer, *Biomaterials*, **24**, 1541-48 (2003).

[28]H. Takadama, M. Hashimoto, M. Mizuno, K. Ishikawa, and T. Kokubo, Newly Improved Simulated Body Fluid, *Key Eng. Mat.*, **254-256**, 115-18 (2004).

[29]P. Layrolle, K. de Groot, J. D. de Bruijn, C. A. van Blitterswijk, and Y. Huipin, Method for Coating Medical Implants. US Patent No: 6,207,218. March 27, 2001.

[30]G. Vereecke and J. Lemaitre, Calculation of the Solubility Diagrams in the System $Ca(OH)_2$-H_3PO_4-KOH-HNO_3-CO_2-H_2O, *J. Cryst. Growth*, **104**, 820-32 (1990).

[31]H. B. Wen, J. G. C. Wolke, J. R. de Wijn, Q. Liu, F. Z. Cui, and K. de Groot, Fast precipitation of Calcium Phosphate Layers on Titanium Induced by Simple Chemical Treatments, *Biomaterials*, **18**, 1471-78 (1997).

[32]J. Choi, D. Bogdanski, M. Koeller, S. A. Esenwein, D. Mueller, G. Muhr, and M. Epple, Calcium Phosphate Coating of Nickel-Titanium Shape Memory Alloys. Coating procedure and adherence of leukocytes and platelets, *Biomaterials*, **24**, 3689-96 (2003).

[33]X. Yin and M. J. Stott, Biological Calcium Phosphates and Posner's Cluster, *J. Chem. Phys.*, **118**, 3717-23 (2003).

[34]A. S. Posner and F. Betts, Synthetic Amorphous Calcium Phosphate and Its Relation to Bone Mineral structure, *Acc. Chem. Res.*, **8**, 273-81 (1975).

[35]P. J. Li, I. Kangasniemi, K. DeGroot, and T. Kokubo, Bonelike Hydroxyapatite Induction by a Gel-derived Titania on a Titanium Substrate, *J. Am. Ceram. Soc.*, **77**, 1307-12 (1994).

[36]K. Onuma and A. Ito, Cluster Growth Model for Hydroxyapatite, *Chem. Mater.*, **10**, 3346-51 (1998).

[37]G. Bauer, Resorptive Bone Ceramic on the Basis of Tricalcium Phosphate. U.S. Patent No. 5,141,511, August 25, 1992.

[38]N. Sasaki and S. Odajima, Stress-Strain Curve and Young's Modulus of a Collagen Molecule as Determined by the X-ray Diffraction Technique, *J. Biomech.*, **29**, 655-58 (1996).

[39]L. Vitagliano, G. Nemethy, A. Zagari, and H.A. Scheraga, Structure of the Type-I Collagen Molecule based on Conformational Energy Computations – The Triple-stranded Helix and the N-terminal Telopeptide, *J. Mol. Biol.*, **247**, 69-80 (1995).

[40]S. H. Rhee and J. Tanaka, Hydroxyapatite Coating on a Collagen Membrane by a Biomimetic Method, *J. Am. Ceram. Soc.*, **81**, 3029-31 (1998).

[41]D. Lickorish, J. A. M. Ramshaw, J. A. Werkmeister, V. Glattauer, and C. R. Howlett, Collagen-Hydroxyapatite Composite Prepared by Biomimetic Process, *J. Biomed. Mater. Res.*, **68A**, 19-27 (2004).

[42]E. K. Girija, Y. Yokogawa, and F. Nagata, Bone-like Apatite Formation on Collagen Fibrils by Biomimetic Method, *Chem. Lett.*, **7**, 702-3 (2002).

[43]A. Oyane, H. M. Kim, T. Furuya, T. Kokubo, T. Miyazaki, and T. Nakamura, Preparation and Assessment of Revised Simulated Body Fluids, *J. Biomed. Mater. Res.*, **65A**, 188-95 (2003).

CHEMICALLY BONDED BIOCERAMIC CARRIER SYSTEMS FOR DRUG DELIVERY

Leif Hermansson
Doxa AB,
Axel Johanssons gata 4-6, SE-75451 Uppsala, Sweden

ABSTRACT

This paper deals with carrier materials for drug delivery based on chemically bonded ceramics (CBC), and specifically Ca-aluminates (CA). The property profile of these CBC-biomaterials and their microstructures give these materials potential as carriers for drugs. The paper describes in some detail the CA carrier system with regard to the technology and chemistry, the biocompatibility and specifically the microstructure and the related loading possibilities of the drugs in the carrier material. The development of microstructures includes different types of porosity, amount of porosity, pore size and pore channel size, and combination of different porosity structures. Specific surface area measurements (BET) of dried fully hydrated Ca-aluminate yield BET- values of > 400 m^2/g, corresponding to a hydrate size of approximately 25 nm, and pore channel sizes of 1-10 nm, in accordance with values from the high-resolution TEM analysis. Complementary porosity above 10 nm is obtained by partial hydration of the precursor material or excess of water in the hydration step, and pore sizes > 100 nm by inert ceramic fillers with phases of oxides of Ti, Si, Ba or Zr, the latter phases selected in order also to increase strength and radio-opacity of the carrier systems discussed. The carrier material can be applied as a solid or a suspension for different types of intake. The drug carrier can also work as an injectable implant.

INTRODUCTION

Carrier materials for drug delivery of pharmaceuticals are based on a broad range of materials, such as organic polymers, metals and sintered ceramics. General aspects of ceramics for use in drug delivery are given by Ravaglioli [1], Lasserre and Bajpaj [2] and aspects of nanophase ceramics for improved drug delivery by Yang, Sheldon and Webster [3]. This paper deals with carrier materials based on chemically bonded ceramics (CBC), primarily Ca-aluminates (CA) and to some extent Ca-silicates (CS). The property profile of these CBC-biomaterials and specifically their microstructures give these materials potential as carriers for drugs and in drug delivery. The chemically bonded ceramics based on Ca-aluminate and Ca-silicate phases as biomaterials within dentistry and orthopaedics have been presented in several papers including two recent Ph D theses [4-17].

There is a need for a carrier material for drug delivery that exhibits well-controlled microstructures, which lend the carrier material opportunities for selected and well-controlled release of the medicament. Issues regarding how and when the medicament is incorporated, where and when it is released, is the theme of this paper to assure high delivery safety for medicaments with regard to release pattern as well as safety aspects of the loaded carrier from chemical and mechanical aspects. A controlled carrier material meeting the above mentioned criteria must also take account of and control of the setting and curing reactions *in vitro* and *in vivo*, as well as to control the porosity of the finally cured material and use of additives and processing agents to assure an optimal microstructure.

This paper will present the great potential which the chemically bonded ceramics exhibit as carriers for drug delivery due to their unique curing/hardening characteristics and related microstructure and the drug loading characteristics.

MATERIALS AND MAIN REACTIONS

Ca-aluminates comprise double oxides of CaO and Al_2O_3. Several intermediate phases exist and these are - using the cement chemistry abbreviation system - C_3A, $C_{12}A_7$, CA, CA_2 and CA_6, where C=CaO and A=Al_2O_3. For the system CaO-SiO_2, the phases $3CaOSiO_2$ (C_3S) and $2CaOSiO_2$ (C_2S) belong to the hydrating phases.

In water environment, the Ca-aluminate and the Ca-silicate cements react in an acid-base reaction to form hydrates. The reactions are temperature dependant. The reaction steps are:

* starting dissolution of CBC into the liquid
* formation of ions, and
* repeated precipitation of nanocrystals (hydrates)
  - Katoite, $3CaOAl_2O_36H_2O$ (C_3AH_6), and Gibbsite $Al(OH)_3$ (AH_3) for the CA-system
  - Poorly crystallized Tobermorite (CSH) and amorphous phases in the CS-system

Data in this paper refer primarily to the CA-phase. The main reaction for the mono Ca-aluminate phase is shown below (H=H_2O):

$$3CA + 12H \rightarrow C_3AH_6 + 2AH_3 \qquad (1)$$

The main reaction involves precipitation on contact areas and within the material, and repeated precipitation occurs until the Ca-aluminate or the water is consumed, resulting in closing of cavities, gaps and voids. This opens up for multipurpose use in many different applications as a biomaterial and as a carrier material for drugs as will be demonstrated in this paper.

Complementary reactions occur, when the Ca-aluminate is in contact with tissue containing body liquid. Several mechanisms have been identified, which control how the Ca-aluminate material is integrated onto tissue. These mechanisms affect the integration differently depending on what type of tissue the biomaterial is in contact with, and in what state (un-hydrated or hydrated) the CA is introduced. These complementary reactions have been described in detail elsewhere [18].

METHODS

The CA-materials have been evaluated comprehensively concerning their bio-compatibility and toxicological endpoints as referred in the harmonized standard ISO 10993:2003 [19], which comprises the following sections:
Cytotoxicity (ISO10993-5), Sensitization (ISO10993-10), Irritation/Intracutaneous reactivity (ISO10993-10), Systemic toxicity (ISO10993-11), Sub-acute, sub-chronic and chronic toxicity (ISO10993-11), Genotoxicity (ISO10993-3), Implantation (ISO10993-6), Carcinogenicity (ISO10993-3) and Hemocompatibility (ISO10993-4).

The corrosion resistance test – using a water jet impinging technique – has been conducted according to EN 29917:1994/ISO 9917:1991, where removal of material is expressed as a height reduction using 0.1 M lactic acid as solution, pH 2.7. The duration time of the test is 24 h. The test starts after 24 h hydration. The test probe accuracy is 0.01 mm. Values below 0.05 mm per 24 h solution impinging are judged as acid resistant [20].

Studies were further complemented by evaluating the chemical reactions and microstructure developed in the CA biomaterial and tissue at the highest level by transmission electron microscopy (TEM) in combination with focused ion beam microscopy (FIB) for site-specific high accuracy preparation described in detail in [21]. Phase and elemental analyses were conducted using traditional XRD, HRTEM, XPS and STEM with EDX.

RESULTS AND DISCUSSION

This paper will to some extent treat and discuss the following topics of high importance when selecting materials to be used as carriers for drug delivery, namely; the chemical composition of the ceramic carrier material, the microstructure including porosity of the carrier material, the optional use of inert additives, the type of medicament, and safety aspects. The great freedom to select different ways of introducing drugs into the carrier system, and how and when release of the drugs can be executed, is the topic of the discussion in relation to results obtainable in the chemically bonded ceramic carrier system. The biocompatibility and bioactivity are presented in detail elsewhere [22]. A good biocompatibility of the CA system was stated.

General aspects

The precursor powder cures as a result of hydration reactions, between the ceramic oxide powder, primarily Ca-aluminates and/or Ca-silicates, and water. Through the hydration, new phases of hydrates are formed (crystalline and/or amorphous ones), which to a great part establish the microstructures needed to control the release of the drug incorporated in the material. The hydration mechanism of these systems involves a reaction where the total volume of the precursor materials and the water (solution) is reduced. This allows a carrier to exhibit open porosity throughout its body even if the total porosity achieved is as low as approximately 5-10 % - a unique feature. The water to cement ratio may be in the interval of 0.3-0.8. A ratio in the interval 0.4 – 0.5 is near complete hydration of the CA material without any excess of water. Excess or limited amounts of water favors complementary porosity, as does hydration at high relative humidity.

The setting time should be relatively short, below 30 minutes, and suitably in the interval of 5-15 minutes. The curing time and temperature are selected to produce controlled microstructure. The carrier materials are suitably hydrated at a temperature above 30°C, since this yields the stable hydrates in the material, and thus a more stable material. The curing before loading and/or before introduction of the material into the body can be done in water or in an environment with high relative humidity (> 70 %). The setting and curing times and temperatures are of specific relevance when the carrier also works as an implant material.

Protection of the carrier and drug during passage through the body or life time in the body can be improved by coating of the precursor material. The coating may suitably be an acid-resistant or a hydrophobic layer. For medical agents sensitive to pH, the pH should be controlled in order to maintain their activity. A suitable pH is in the interval of 6-8. This can be achieved by introduction of a buffer. The buffer may favorably be a biocompatible one based on hydrogen-phosphates.

The CBC selected yields by itself a radio-opacity of approximately 1.5 mm. In order to impart a higher radio-opacity of the carrier, phases with a high electron density are added. This allows the drug to be located in the body using X-ray techniques. Examples of such phases are ZrO_2 and Sr- and Ba-containing glasses. The introduction of these phases also strengthens the materials.

The microstructure developed consists of nano-size hydrates and nano-size channels, located between said hydrates, having a size of about 10-50 nm and 1-5 nm, respectively. Complementary porosity above 10 nm can be achieved by 1) partial hydration of the precursor material, 2) excess of water in the hydration step and/or 3) additional porous inert fillers, additional ceramics (such as hard particles or other hydrated phases) and other porous materials such as stable polymers and stable metals. The pore size can thus be controlled in a carrier from 1-2 nm to micrometer size level, typically < 50 micrometer. The Fig. 1 shows the actual nanostructure of the hydrated material and Fig. 2 a HRTEM picture of the same [13,22].

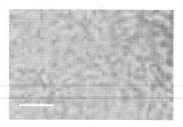

Fig.1. The hydrated area of a Ca-aluminate based material (white bar = 50 nm)

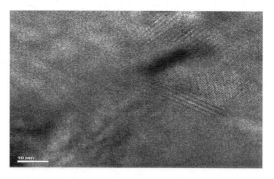

Fig. 2 HRTEM showing the nanostructure of the hydrated CA-phase, white bar = 10 nm

The property profile of the CA-material can be seen in Table I, which summarizes the typical interval for each property. The intervals presented are mainly related to the completeness of the hydration, the water to cement ratio, and the type of filler particles introduced.

Table I. Typical property values, and interval for properties of CA-material based systems

Property	Typical value	Interval
Compression strength, MPa	150	60-270
Young's modulus, GPa	15	10-20
Thermal conductivity, W/mK	0.8	0.7-0.9
Thermal expansion, ppm/K	9.5	9-10
Flexural strength, MPa	50	20-80
Fracture toughness, MPam$^{1/2}$	0.5	0.3-0.8
Corrosion resistance, water jet impinging, reduction in mm	< 0.01	-
Radio-opacity, mm	1.5	1.4-2.5
Process temperature, °C	> 30	30-70
Working time, min	3	< 4
Setting time, min	5	4-7
Curing time, min	20	10-60
Porosity after final hydration, %	15	5-60

Drug loading and manufacturing aspects

The loading of the drug can be performed in several ways. The drug may be included, either partially or fully, in the powder or in the hydration liquid with or without any processing agents. The powder may be composed of non-hydrated or hydrated ceramics, porous additives such as sintered ceramics, stable polymers or metals. The drug may be included in one or more of these powders or the liquid, and may be mixed with or incorporated into any open porosity of the components.

Time and temperature for hydration are selected with regard to the drug and drug loading and to the selected release criteria. Temperature, as well as type of precursor powder, amount of precursor powder and processing agents, control the time selected for manufacturing the carrier. The manufacturing of the carrier can be done completely before or during loading of the drug. This renders a controlled release time to be selected from a few hours to days and months.

The drug is introduced in the carrier by mixing the drug into the precursor powder, or the hydrated CBCs or other porous phases. The material can be formed into a paste by mixing it with a water-based hydration liquid. The powder can also be pressed into pellets, which thereafter are soaked in the liquid. The paste or the soaked pellets start to develop the microstructure that to a great extent will contribute to the controlled release of the drug. The time and temperature after the mixing will determine the degree of hydration, i.e. the porosity obtained. The porosity can be controlled within a broad interval of open porosity.

The drugs can be loaded in the water-liquid, in the pore system of inert filler particles and in processing agents (accelerators, retarders, viscosity controlling agents and other rheological agents). Thus drugs can be loaded both during formation of hydrates or after hydration by infiltration. The infiltration comprises water-penetration of precursor materials or hydrated materials using wetting at normal pressure, during vacuum, or overpressure. For hydrophobic medical agents, the agent can be easily mixed into the precursor powder or together with the ceramic or other filler materials.

The amount of drug loaded in the carrier is determined by the content of the drug in the dry powder and the hydration liquid. The liquids involved in the introduction of the drug into either the dry powder or the hydration liquid are easily controlled in charged amount of liquids. The reaction takes preferentially place in high humidity, where no liquid in the carrier is vanished into the environment.

Optionally, the carrier powder may comprise inert oxides of Ti, Si, Ba or Zr, in order to increase strength or radio-opacity. The oxides may take the form of porous and/or dense particles. The incorporation of the drug or medical agent into the carrier material in the porous inert ceramic material, may be performed by filling the pores of the inert ceramic with the drug, by mixing it with the powder prior to mixing it with the hydration liquid, or mixing it with the hydration liquid prior to mixing it with the precursor powder. Depending on the type of drug delivery for which the carrier material is intended, a combination of one or more of these techniques may be used.

The carrier material may further comprise a third type of ceramics, including one or more of other hydrated or non-hydrated hydraulic phases, such as calcium phosphates, calcium sulphates, as well as hydroxyapatite. The carrier material may further comprise a forth inert material of a porous polymer or porous metal. The carrier loading capacity is estimated to be below 0.5 v-% to as high as approximately 10 v-%.

Drug release control aspects

The following properties are of significance with regard to the carrier for controlling the drug release; Type of ceramic precursor for producing the chemically bonded ceramic, grain size distribution of the precursor powder particles and general microstructure of the material, the microstructure of the additional particles for drug incorporation, and additives to ensure complementary porosity.

Type of chemically bonded ceramic

The preferred chemical compositions, with an inherent property profile to meet the features described in this presentation, are those based on chemically bonded ceramics, which during hydration consume controlled amount of water. The preferred systems available are those based on aluminates and silicates, which both consume a great amount of water. Phases such CA_2, CA, C_3A and $C_{12}A_7$, and C_2S and C_3S in crystalline or amorphous state (C= CaO, A $=Al_2O_3$, S = SiO_2) may be used. The pure aluminate and silicate phases are not available on the market. These were synthesized at temperatures close to 1400 °C for 3-5 hrs. The Ca-aluminate and Ca-silicate phases may be used as separate phases or as mixtures of phases. The above mentioned phases, all in non-hydrated form, act as the binder phase (the cement) in the carrier material when hydrated.

Grain size distribution

The grain size of the precursor powder particles should be below 20 μm - this to enhance hydration. The precursor material is transformed into a nano-size microstructure during the hydration. This reaction involves dissolution of the precursor material and repeated subsequent precipitation of nano-size hydrates in the water (solution) and upon remaining non-hydrated precursor material. This reaction favorably continues until all precursor materials have been transformed, or to a porosity determined by partial hydration using the selected time and temperature.

General microstructure of material

The paper presents a couple of unique reaction conditions related to the production of materials having a variety of microstructures. These include development of microstructures having different 1) type of porosity, 2) amount of porosity, 3) pore size and pore channel size, and 4) combination of different porosity structures.

Porosity generated during the hydration of the Ca-aluminates and Ca-silicates is open porosity due to the reaction mechanism, and can be in the interval of 5-60 vol.-%. The average pore channel size (i.e. the diameter of the pores formed between the particles of the hydrated material) may be 1-10 nm. The crystal size of the reacted hydrates is approximately 10-50 nm. This was established by BET-measurements, where the specific surface area of dried hydrated CA was determined to be in the interval 400-500 m^2, corresponding to a particle size of approximately 25 nm. See also Fig. 1. When short hydration time and/or low amount of water, or moisture at relative humidity > 70 %, are used, additional porosity is achieved with pore sizes in the interval 0.1-1 micrometer due to incomplete reaction. See also Fig. 3 below.

Microstructure of additional particles (additives) for drug incorporation

The microstructure of the complementary additives which are penetrated by/loaded with the active medical agent is primarily characterized by its porosity, which should be an open porosity in the interval of 15-70 volume-%. The average pore size determined by Hg-porosimetry is in the interval of 0.1-10 μm. This is a complementary additive microstructure to that of the main structure based on the chemically bonded ceramics. Examples of such additives include inert and hard ceramics such as oxides and/or carbides and/or nitrides. These phases yield a carrier material having increased strength and chemical resistance. The additional particles with a pore size in the interval 0.1-10 micrometer are introduced to speed up the release rate from slow release down to release time of a few hours (< 5 hrs) and can favorably be used to be loaded with additional drugs for rapid release. The figure 3 summarizes the pore size intervals obtainable with different ceramic used for drug delivery.

Log (pore size in nm)

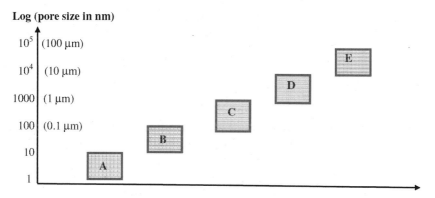

Fig. 3 Pore size interval for different types of ceramics used as drug carriers
(A = Fully hydrated CBC, B = Partially hydrated CBC, C = Sintered ceramics (submicron) ,
D = Sintered ceramics (partially or coarse grain), and E = Others

The amount of drug released in a specific case is controlled by the amount of the drug in the carrier, and how the carrier material is designed. Using the mixed powder cement and an inert phase opens up for use of combined drugs, e.g. for rapid release based on the inert phase, medium release rate based on partially hydrated phases, and slow release based on fully hydrated cement phases. Test materials have been used with loading of the drugs in the interval 1-10 mg/100 mg carrier material. The figure 4 shows schematically release rates using corresponding selected pore structures of the carrier material. The material types C, B, and A from Fig. 3 were used.

Amount released (mg drug/ 100mg carrier material)

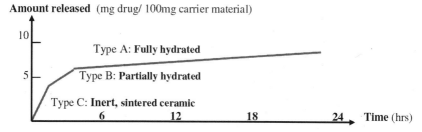

Fig. 4 Schematic release of drugs using different pore structures,

Pharmaceutical compositions
      The composition can be in the form of a solid or a suspension for different kinds of intake from oral intake to percutaneous injection.

The medicament can be of any kind. Preferable medicaments are those chosen from for cancer/tumor treatment, vascular treatment, bone restoration, antibacterial and anti-inflammatory agents, pain relief drugs, anti-phlogistics, drugs antifungal agents, antivirus agents, analgesics, anticonvulsants, bronchodilators antidepressants, auto-immune disorder and immunological disease agents, hormonal agents, TGB-beta, morphogenetic protein, trypsin-inhibitor, osteocalcine, calcium-binding proteins (BMP), growth factors, Bis-phosphonates, vitamins, hyperlipidemia agents sympathetic nervous stimulants oral diabetes therapeutic drugs oral carcinostatics, contrast materials, radio-pharmaceuticals, peptides, enzymes, vaccines and mineral trace elements or other specific anti-disease agents.

Non-active ingredients can be added. With non-active ingredients is meant water, alcohol, thickening agents, sweeteners, colours, antioxidants or other additives which may be useful for stabilizing the composition.

CONCLUSION

- The ceramic carrier chemistry presented in this paper allows for loading of almost any medicament. The drugs can favorably be loaded in the water-liquid, in the pore system of inert filler particles and in processing agents. Thus drugs can be loaded both during formation of hydrates or after hydration by infiltration. For hydrophobic medical agents, the agent can be easily mixed into the precursor powder or together with the second ceramic filler.
- The carrier material system described exhibits well-controlled microstructures on the nano-size level, which lends the carrier material opportunities for selected and controlled release of the medicament. The release time is controlled mainly by the contents of the hydrated chemically bonded cement phases, the higher the content of the cement, the longer the release time. The longest release time is achieved for fully hydrated phases with a water content close to the w/c required for complete hydration of the precursor Ca-aluminate.
- By introducing optional additives, or by changing the w/c ratio, the release time can be controlled from a few hours to more than one day. The release time is also dependant upon where the drug is placed. In cortical bone a release time of months seems possible.
- The carrier may be used as a vehicle for transport and delivery of the medicament as an injectable implant. The combination of the material as carrier and implant material makes site-specific placement of drugs and implants possible.

ACKNOWLEDGEMENT
The author expresses his great gratitude to all personnel at Doxa AB for input under a ten-year period.

REFERENCES
[1] Ravaglioli et al. *J Mater Sci Mater Med.* 2000 11(12):763-7
[2] Lasserre and Bajpaj, Critical Reviews in Therapeutic Drug Carrier Systems, 15,1 (1998).
[3] L. Kraft, Calcium Aluminate Based Cement as Dental Restorative Materials, *Ph D Thesis,* Faculty of Science and Technology, Uppsala University, Sweden. 2002
[4] L Yang, B Sheldon and T J Webster, Nanophase ceramics for improved drug delivery, *Am. Ceram Soc. Bulletin* Vol 80, [2], 24-31 (2010)
[5] J. Lööf, Calcium Aluminate as Biomaterial – Synthesis, Design and Evaluation, *Ph D Thesis,* Uppsala University, 2008

[6]L. Kraft and L. Hermansson, Hardness and dimensional stability of a bioceramic dental filling material based on calcium aluminate cement, *Am. Ceram. Soc., Advanced ceramics, Materials and structures*, Vol 23B, [4] (2002)

[7]H. Engqvist, J-E. Schultz-Walz, J Lööf, G.A. Bottom, D. Mayer, M.W. Phaneuf, N-O. Ahnfelt, L. Hermansson, *Biomaterials* Vol 25, 2781-2787 [2004]

[8]H. Engqvist, M. Couillard, G.A. Botton. M.P. Phaneuf, N. Axén, N-O Ahnfelt and L. Hermansson, In vivo bioactivity of a novel mineral based based orthopaedic biocement, *Trends in Biomaterials and Artificial Organs*, Vol 19, 27-32 (2005)

[9]L. Hermansson, U. Höglund, E. Olaisson, P. Thomsen, H. Engqvist, Comparative study of the bevaiour of a novel injectable bioceramic in sheep vertebrae, *Trends in Biomater. Artif. Organs*, Vol 22, 134-139 (2008)

[10]A. Faris, H. Engqvist, J. Lööf, M. Ottosson, L. Hermansson, In vitro bioactivity of injectable ceramic orthopaedic cements, *Key Eng. Mater.* 309-11, 1401-1404 (2006)

[11]Y. Liu et al, Aspects of Biocompatibility and Chemical Stability of Calcium-Aluminate-Hydrate Based Dental Restorative Material, Paper IX in *Ph D Thesis* by L. Kraft, Uppsala University (2002)

[12] J. Lööf, A. Faris, L. Hermansson, H. Engqist, In vitro mechanical testing of two injectable materials for vertebroplasty in different synthetic bone, *Key Eng. Mater.* Vol 361-363, 369-372 (2008)

[13] L. Hermansson, A. Faris, G. Gòmez-Ortega and J. Lööf, Aspects of dental applications based on materials in the system $CaO-Al_2O_3-P_2O_5-H_2O$, *Ceramic Eng. and Sci. Proc.*, Volume 30, 71-80

[14] J. Lööf, H. Engqvist, K. Lindqvist, N-O.. Ahnfelt and L. Hermansson, Mechanical properties of a permanent dental restorative material based on calcium aliminte, *J Mater. Sci., Materials in Medicine*, Vol 14, 1033-1037 (2003)

[15] H. Engqvist, S. Edlund, G. Gómez-Ortega, J. Lööf, L Hermansson, In vitro mechanical properties of a calcium silicate based bone void filler, *Key Eng. Mater.* 309-311, 829-832 (2006)

[16] L. Hermansson, L Kraft and H Engqvist, Chemically bonded ceramics as biomaterials, *Key Eng. Mater.* Vol 247, 437-442 (2003)

[17]H. Engqvist, J. Lööf, L. Kraft, L. Hermansson, Apatite formation on a biomaterial-based dental filling material, *Ceramic Transactions,* Vol 164 , Biomaterials: Materials and Applications V, (2004)

[18]L. Hermansson, J. Lööf and T. Jarmar, Integration mechanisms towards hard tissue of Ca-aluminate based materials, *Key Eng. Mater.* Vol 396-398, 183-186 (2009)

[19] ISO 10993:2003

[20]EN 29917:1994/ISO 9917: 1991

[21] H. Engqvist, G. A. Botton, M. Couillard, S Mohammadi, J. Malmström, L. Emanuelsson, L Hermansson, M. W Phaneuf, P Thomsen, A novel tool for high-resolution TEM, *J Biomed. Mater. Res.* Vol49, 257-281 (2006)

[22] L.. Hermansson, A. Faris, G. Gómez-Ortega and J. Lööf, Biocompatibility Aspects of dental applications based on materials in the system $CaO-Al_2O_3-SiO_2-P_2O_5$, *Ceramic Eng. and Sci. Proc.*, Volume 30, 59-70 (2009)

Porous Ceramics

LOW-O2 TECHNOLOGY FOR THERMAL TREATMENT OF HIGH QUALITY POROUS
CERAMICS

Hartmut Weber
Riedhammer GmbH

This lecture considers the key process step of debindering, presents a newly developed
debindering technology for porous ceramics and explains the application of this technology
for intermittent as well as for continuous production kilns.

Porous ceramics are used in large quantities for numerous applications. One of the highest
volume examples of this is as diesel particulate filters (DPF) in the automobile industry: The
DPF is installed in the exhaust gas flow. During the filtering process the exhaust gas flows
through the porous ceramic walls and the sooty particles get stuck there. With increasing
sooting of the filter, the pressure drop increases. Therefore it is necessary to provide for an
efficient regeneration of the filter through the effective combustion of the accumulated soot at
regular intervals. During this regeneration process, the porous ceramics are exposed to a
severe thermal shock. Porous ceramic honeycomb structures based on SiC, see **Figure 1** and
lately also of aluminium titanate (ATI), see **Figure 2** have been proven to be adequate in this
connection.

Figure 1
Typical design of SiC-filter

Figure 2
Typical design of ATI-filter

The shaping of the DPF honeycomb structures is usually made by continuous extrusion.
During the extrusion process organic shaping additives of up to 25 % by weight are added to
the body to make it more plastic. These additives must be removed completely before the
sintering process, as carbon residues left in the product can negatively affect both the sintering
process and the finished characteristics of the DPF. The process step of removing these
organics is commonly called „Debindering". For mass production of DPFs, thermal
debindering at atmospheric oxygen concentrations is the common practice.

The process control during the thermal debindering is critical because the breakdown velocity
of the polymers must not exceed the evacuation velocity of the pyrolysis products. Otherwise
pressure would build up inside the body, leading to cracks in and corrosion of the component.
During the debindering process with an air atmosphere, very distinct exothermic and
endothermic reactions take place, see **Figure 3**.

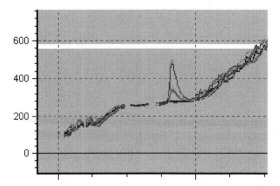

Figure 3
Severe partial temperature increase inside part body due to exothermic reactions

To deal with these, special process controls must be implemented so the heat treatment acts homogeneously on the whole component. This process control allows for very small rates of temperature rise during thermal debindering, which results in long kiln times. The resulting debindering process can take 50% and more of the total kiln time (Debindering +Sintering) and must be carried out in special ovens. After the debindering process, the DPFs are transferred into the actual sintering kiln which means an additional handling effort for the serial production that cannot be cost-effective in the long run. Many times, to minimize the defects from handling of the parts, a pre-sintering process is carried out immediately after the debindering, but prior to transfer increasing the energy required to process the component.

Recognizing this shortcoming, an investigation using the laboratories of RIEDHAMMER in Nuremburg, Germany and the laboratories of our affiliated companies in the TEAM by SACMI family, was undertaken. Samples of a typical DPF body were tested through the debindering temperature rage with thermogravimetric analysis (TGA) and differential thermal analysis (DTA) it has been found that the debindering process takes place in a temperature range between 150-500°C, see **Figure 4**. During this process in an air atmosphere very distinct partial exothermic and endothermic reactions take place. By reducing the oxygen content in the furnace atmosphere, the intensity of these reactions and thus the amplitude of the thermal fluctuations occurring in the component could be reduced. Tests with varied heating rates showed that the heating rate does have an influence on the intensity of the exothermic and endothermic reactions taking place, but that this influence is not as important as that of the oxygen content, see **Figure 5**.

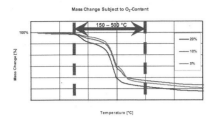

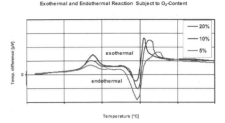

Figure 4
Differential Thermal Analysis

Figure 5
Exo- /Endothermal reactions subject to
O2-Content

Based on the results from these laboratory tests, a pilot-sized kiln at the RIEDHAMMER laboratory was built, allowing small lots of full-size samples to be tested. With this, it was proven that specific control of the oxygen content in the kiln atmosphere during the debindering process which allows the intensity of the endothermic and exothermic reactions taking place as well as the amplitude of the thermal fluctuations occurring on the component to be controlled had a significant impact on the processing cost of a DPF. This technology makes it possible to remove all organic substances contained from the green bodies with the maximum possible physical and chemical debindering rate without quality loss, significantly reducing the required debindering period, see **Figure 6**.

To affect this in full-scale, proprietary schemes for atmosphere and temperature control in combination with a continuous, recirculating kiln atmosphere during the debindering process were developed. Separate innovations were developed to allow for an automatic change to direct atmosphere control by means of gas and air regulation of the burners installed in the kiln at higher temperatures. The further heating and process control according to the sintering curve may follow immediately. This eliminates the usual time-consuming, defect-creating and costly steps of pre-sintering to 900°C, cooling down and subsequent transfer of the DPFs.

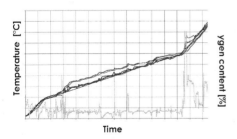

Figure 6
Temperature profile during debindering
with LOW-O2-Technology

Figure 7
Periodic kiln with LOW-O2-
Technology

Based on the results of this research, the technology has been applied at several full-size DPF manufacturing plants. The full-size system is marketed as "LOW-O2-Technology by RIEDHAMMER", and has successfully been integrated as part of new periodic and continuous kiln plants. Results predicted from the pilot plant tests have been demonstrated in a mass-production environment. Owners of these systems report that they were able to achieve significantly higher yield in a shorter processing time and at a lower capital and energy cost than possible with the two-stage production with separate debindering and sintering kilns.

RETICULATED SIC FOAM X-RAY CT, MESHING, AND SIMULATION

Alberto Ortona[*]. Simone Pusterla[*], Solène Valton[§]

* ICIMSI SUPSI Galleria 2, 6928 Manno-CH
§ RX Solutions 11 Route de la Salle, 74960 Cran-Gevrier-F

ABSTRACT

Silicon carbide ceramic foams are currently employed in applications such as porous burners and heat exchangers where high thermal loads are applied. These loads can be extremely severe and, up to date; the only reliable routine to assess components performances is the time and money consuming "trial and error" method. Furthermore, due to the irregular foam shape, tests must be repeated several times for statistical reasons.

This paper presents a methodology to simulate ceramic foams effective thermal conductivities from 3D meshes built up from CT output data. A first commercial code combines image filtering and segmentation followed by object surface meshing. Further meshing refinement and volumetric meshing is done with another software.

Meshes are than imported into a FEA code and thermal loads applied allowing to simulate heat fluxes through foam ligaments. Foam effective thermal conductivity is then calculated and results compared with values obtained with experimental data and literature.

INTRODUCTION

Rigid reticulated foams are made from a variety of materials, namely metals, carbons, and ceramics. They can be applied in many fields of advanced manufacturing technologies [1.]. Open reticulated foams are characterized by a network of interconnected ligaments capable of continuous thermal conduction. Thanks also to their high specific surface area they can enhance heat-transfer via radiation and convection. Open-cellular foams have recently seen increasing their applications to enhance heat transfer in a variety of thermal devices [2.][3.][4.].

Beside analytical models, which predict with different approximations the effective thermal conductivity of porous materials [5.][6.][7.], FE simulation has become a well consolidated tool for materials properties prediction under thermal loads. Looking at ceramic reticulated foams, the generation of FE computational models is strictly linked to CAD tools [18.]. In case of unmappable geometries this approach can no longer be applied because of the impossibility to describe these surfaces with mathematical functions.

The use of X-Ray Computed Tomography in material analysis makes possible the description and simulation of complex geometries. Computed Tomography is an imaging method employing tomography (imaging by sections or sectioning, through the use X rays) created by computer processing. Digital geometry processing is used to generate a three-dimensional representation of the object (outside and inside) from a large series of two-dimensional X rays projections taken while the object is rotating around a single axis.

X-ray images, commonly referred to as radiographies, are transmission images: an X-ray source produces a beam of X-rays in a given direction, the beam first meets the object to be scanned and the transmitted beam is then collected by a detector. In a first approximation, radiographies, also called projections, can give qualitative object information through a coefficient that characterizes matter. This coefficient is known as attenuation coefficient and depends on the electronic density of matter and the energy of point in space.

The purpose of X-ray computed tomography (CT) is to determine the attenuation coefficient ($\mu(\vec{x})$) at any point in the object. It was first discovered by Hounsfield in the late sixties. Tomographic reconstruction mathematical foundations rely on the Radon transform [11.]. It allows the reconstitution of $\mu(\vec{x})$ from a large number of radiographies taken around an object. Different algorithms and reconstruction techniques exist [13.][14.], as well as different acquisition geometries

The result of 3D tomographic reconstruction is a digital volume where each elementary unit called "voxel" represents an approximation of the attenuation coefficient in the object at that position. Tomographic volumes are conveniently stored on computers as stacks of pictures which correspond to parallel slices across the volume. The voxel size determines the resolution of the CT reconstruction

Aim of this paper is to develop and validate a computational method which reproduces the thermal behaviour of real reticulated foam allowing to understand better which factors affect its effective thermal conductivity .

EXPERIMENTAL PROCEDURE

Materials

ErbisicR foams (Erbicol SA Balerna, Switzerland) were employed for the experiments. Manufacturing of ceramic foams is based on commercially available PU foams which were used as templates [12.]. The process starts with an α-SiC powder slurry which is applied as a dip coating on the polymer network (ligaments), followed by pyrolysis. By controlling the first part of this process, staring from the same template, it is possible to obtain different coating thicknesses and thus different porosities. Two ErbisicR foams named ErbisicR_H and ErbisicR_L (Figure 1) were produced following the above mentioned procedure and subsequently machined with diamond tools to (20x20x20 mm^3) dimensions.

To improve the strength of the ligaments ceramic network, a second polymer infiltration and pyrolysis was also performed. Finally, silicon melt infiltration was used to obtain a reaction bonded β-SiC-silicon matrix binding the dispersed α-SiC powder. The final product is a three-dimensional network of Si-SiC ligaments (bulk density 2.83 g/cm^3, measured via He picnometry). The extra silicon content of as-received ErbisicR foams is approx. 30% by weight [10.]. This Silicon fills intra- SiC particle spaces and hollow ligaments, respectively on a microscopic and on a macroscopic scale. Pore sizes are about 4-7 mm and are equivalent to those of commercial reticulated foams with a pore size of about 10 Pores per inch (PPI).

ErbisicR _H   ErbisicR _L

Figure 1 ERBSIC R foams

Dimensions tolerances were kept quite tight for object dimensions   calibration after XTC reconstruction. Table 1 summarizes average physical sample properties.

Sample ID	L1 [mm]	L2 [mm]	L3 [mm]	Volume [mm3]	Mass [g]	Density [g/cm3]	Porosity
ErbisicR_H	20.09	19.99	20.16	8093.00	3.60	0.44	0.84
ErbisicR_L	20.10	20.01	20.01	8047.15	2.47	0.31	0.89

Table 1 ErbisicR foams characteristics

X-Ray Computed Tomography

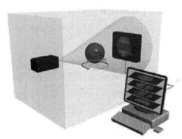

Figure 2 Principle of tomographic acquisition

CT data used in this paper were acquired by RX Solutions on a laboratory microCT "EasyTom130" (RX Solutions, Cran Gevrier - France).

It contains a Thales Hawkeye (France) 130 kV compact X-ray unit. The unit includes a high voltage electrical generator and a tube with a tungsten target which emits X-ray photons when hit by high energy electrons (mainly by "bremsstrahlung"). The energy (and thus the wavelength) of the photons directly depends on the voltage of the generator. The user can set the voltage ranging from 30 to 130 kV depending on the sample to be analysed. Low energy photons will provide more contrast in low density material (plastic, tissues), while high energy photons are requested to inspect high density material (metals, minerals). The X-ray photons are emitted in a divergent beam by a focal spot which has to be smaller than the size of the reconstructed voxels.

The X-ray detector is a flat panel from Varian (Paolo Alto, CA, USA). The first layer of the panel is a conversion screen made of crystal scintillator (CsI in this case) which will convert X-ray photon to visible light. Amorphous silicon receptor and high speed electronics acquire up to 30 digital radiographies per second. The detector was used in 2x2 binning mode where pixel size is 0.254 mm. The size of the reconstructed voxels depends on the position of the rotating stage supporting the object: it is given by the geometrical magnification of the projection of the object on the detector screen.

Translation axes driven by the user with a joystick allow to easily set the geometry of the system so that the projection of the object covers the whole surface of the detector and thus optimize the magnification factor.

The tomographic acquisition consists in the acquisition of 600 projections over 360° (Figure 2). For each angular position, the registered image is the result of the average of several frames in order to reduce the Poisson noise native on X-ray projections. The sample stage is then rotated and a new projection is acquired. At the end of the acquisition process, the volume is reconstructed via filtered back projection. The conditions of acquisition are summarised in the following table.

Table 2 Main aquistion parameteres

X-ray generator voltage	Detector pixel size	Number of projections	Reconstructed voxel size
100 kV	0.254 mm	600	0.032 mm

A projection is shown on Figure 3. Reconstructed data consist in a stack of slices across the sample. Dedicated software can then be used to display slices in x, y, and z direction, as well as a 3D surface rendering of the reconstructed volume (Figure 5). The grey level of the pixel is proportional to an approximation of the attenuation coefficient of matter. Black pixels represent minimum attenuation value (i.e. air) and dark to light grey represent matter with increasing value of electronic density.

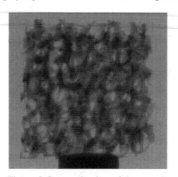

Figure 3 One projection of the sample

Volume reconstruction

XCT output 2D images (Figure 4-a) were visualized and further processed with commercial software Avizo (VSG, Visualization Science Group, Inc. Burlington, MA USA). For the present work, an equalization filter was used (Figure 4-b).

The image segmentation process consisted in a separation of two different domains, meaning in this case, the ceramic foam and the external environment (Figure 4-b). During this operation differences (see Figure 4-a) between Si, (dark grey) and SiC (bright grey) were lost. It means that Si-SiC and extra Silicon, filling hollow struts, were assumed as one material.

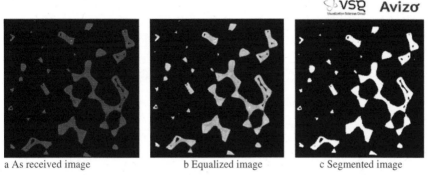

a As received image   b Equalized image   c Segmented image

Figure 4 Image processing

The software then reconstructed a three dimensional interfacial isosurface mesh of the two foams (Figure 5), composed of first order triangular elements.

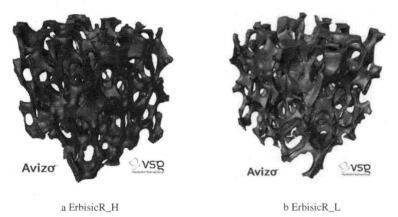

a ErbisicR_H   b ErbisicR_L

Figure 5 3D rendering of ErbisicR foams

The number of elements of the first meshes exceed 6 million (Figure 6-a). To reduce their number a mesh coarsening algorithm was applied. Mesh coarsening was performed in a recursive adaptive way, (i.e. coplanar elements were successively combined to form bigger ones). This process lead to a lower quality mesh with sharper elements (Figure 6-b).

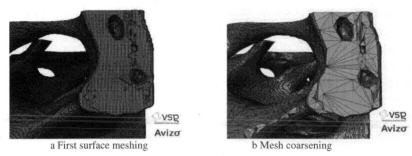

| a First surface meshing | b Mesh coarsening |

Figure 6 Foam surface mesh reconstruction

The coarsened surface meshes were subsequently processed with the commercial Software Hypermesh (AltairTroy, Michigan, USA). Meshes were first checked to eliminate questionable connected triangles and free edges, as well as sharp triangular elements with angles smaller than 20°. The internal volume was then discretised in linear-4-nodes tetrahedral elements through unstructured meshing [9.].

For each, foam two dummy blocks were then generated on two opposite foam faces and meshed . To study foam anisotropy due to their elongated cell shape (Figure 7), blocks were placed perpendicular to:

- The elongated cell height H (Figure 8-a)
- The elongated cell width D (Figure 8-)

Blocks provide a simple and dimensionally known surface which allows the overall heat flux calculation [18.].

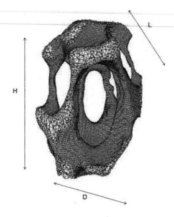

Figure 7 Elongated tetrakaidecahedron repeating unit cell [15.]

To realize a solid geometry comprising foam and blocks, a single external surface mesh was built around the two blocks and the foam. During this process some hollow ligaments, with internal bores smaller than the element size, were closed, virtually decreasing specimen porosity. Meshing lead to four computational domains with different porosity and cells orientation.

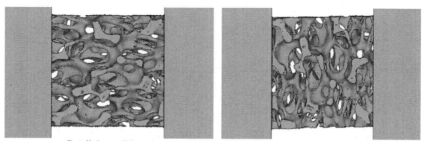

a Parallel to cell length   b Perpendicular to the cell length

Figure 8 Dummy blocks positioning on ErbisicR_H foam.

FE Analysis

Thermal analysis was performed using Marc and Mentat software (MSC Software Corporation, Santa Ana, CA USA). Since ErbisicR bulk material thermal properties were not available, a Si-SiC material with similar microstructure was chosen as reference (INEX Incorporated NY, USA) its thermal conductivity and density are 160W/m K and 2.85 g/cm^3 respectively. A fictitious material with thermal conductivity two order of magnitude greater than SiC was assigned to the external blocks to consider negligible their contribution to the effective heat conductivity.
The steady state heat conduction applied to the model was given by

$$\nabla \cdot (-K_{eff}\nabla T) = 0 \qquad (1)$$

Where K_{eff} denotes the effective thermal conductivity and T the temperature.
The following boundary conditions were applied to the computational domains:
- x=0, T=295K at the left face of the block;
- x=L, T=300K at the right face of the block.
- adiabatic conditions all over the other external surfaces of the computational domain.

Effective thermal conductivity measurements

The experimental equipment shown in Figure 9 was realized for thermal conductivity measurements according to other works [17.]

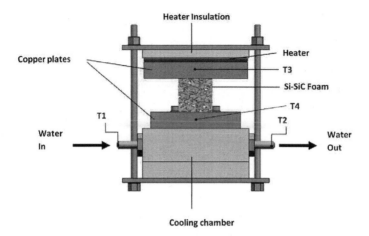

Figure 9 Thermal conductivity measurements set up.

The equipment measured one dimensional heat flow, assuming to reach a steady state. The upper copper plate was heated (393K) with a planar heater while the lower copper plate was cooled by flowing water at ambient temperature.

The temperatures of both Copper plates, as well as the temperature of water at the inlet and outlet of the cooling chamber were measured with four Type K thermocouples. Location of temperature measurements is indicated in Figure 9 as T1, T2, T3, T4.

T3 and T4 thermocouples were positioned in the centre of the copper plates; the temperature of the water was measures pitting T1, T2 thermocouples inside the cooling pipes. Temperatures were measured using a Lab View acquisition system (National Instruments Corp. Austin TX USA) which automatically calculated heat flow and effective thermal conductivity.

High thermal conductivity silicone liquid gap filler TIM-LGF 2000 (TimTronics Yaphank, NY USA), was used to ensure contact between the foam and copper blocks.

The apparatus was further insulated using "rockwool" (K = 0.045 W/mK).

Cooling chamber and the upper insulation were made of fibreglass/epoxy composite material (K= 0.13 W/mK).

In these conditions it was assumed that the heat flowing through the porous medium $\dot{q}$ is equal to the heat transferred to the water. It can be calculated as:

$$\dot{q} = \dot{m} \cdot c_{H_2O} \cdot \Delta T_{H_2O} \qquad (2)$$

Where $\dot{m}$ is the water mass flow rate, c_{H_2O} is the specific heat of the water and ΔT is the temperature difference from the inlet and outlet of the cooling chamber.

Once reached a steady state condition, the effective thermal conductivity of the sample was then calculated as

$$K_{eff} = \frac{\dot{q} \cdot L}{\Delta T_{Cu}} \qquad (3)$$

Where $\dot{q}$ is the heat flux obtained from equation (2) , L the foam thickness and ΔT_{Cu} the temperature difference between the warm and cold copper plates.

Beside measurements on ErbisicR_L and ErbisicR_H in two perpendicular directions as per Figure 8, other ErbisicR specimens with different porosities and pore orientation were tested .

RESULTS

For the simulation it was assumed, in case of steady state heat conduction without internal sources, that the heat flowing trough faces at x =0 and x=L is equal and must be the same of the heat flowing through the foam. Once the FE code calculated the heat flux through one block face, the effective thermal conductivity of the porous media (

Table 3) was calculated as:

$$k_{eff} = q \cdot \frac{\Delta x}{\Delta T} \qquad (4)$$

Where q is the heat flux through the foam, Δx is the foam length L and ΔT the temperature difference applied to the computational domain. L values are different from cube dimensions because they were measured with the meshing code, and because , when intersecting dummy blocks with the foams, some foam parts where cut.

Table 3 Simulation results and calculated K_{eff}

Model	Heat Flux [W/m²]	L [m]	ΔT [°C]	K_{eff} [W/mK]
ErbisicR_H $=$	1686.45	0.0203	5	6.85
ErbisicR_H $\perp$	1353.33	0.0196	5	5.30
ErbisicR_L $=$	1866.66	0.0165	5	6.16
ErbisicR_L $\perp$	1017.50	0.0200	5	4.07

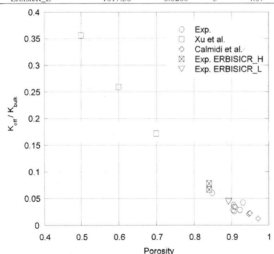

Figure 10 Measured relative thermal conductivity vs. foam porosity for the present and the literature

Figure 10 shows the effective thermal conductivity experimentally measured in this work and the values reported in other works from the literature [17.][18.]. For a better comparison with reticulated foams made of other materials , thermal conductivities values were normalized in respect of their bulk material thermal conductivities.

DISCUSSION

In general, effective thermal conductivity decreases at the increasing of the porosity. This is because at low temperature, the effective conductivity of foam depends mainly by pure conduction since the effect of convection and radiation can be neglected [19.].
Sample with similar porosity show high difference in K_{eff} (Figure 10) or, on the other hand, sample with different porosity can have similar K_{eff}. This is because several factors affect heat conduction apart from porosity, among them is cell orientation.

K_{eff} experimental values calculated in the present work (Figure 10) seems coherent with values from other works, that allowed to compare experimental results with simulation data.
In this regard , K_{eff} for ErbisicR foams, was experimentally measured and calculated via simulation in two perpendicular directions. Output values are reported in Table 4.

Data from simulation are always lower than the experimental values; the difference increases with the decreasing of the porosity. The reason could be the K_{bulk} value utilised as input for simulation, this value, which has been assumed to be the foams one, could be higher for ErbisicR foams.

Table 4 Comparison between experimental and simulated results

Specimen	Porosity	K_{eff} Experimental [W/mK]	K_{eff} Simulation [W/mK]	Difference [W/mK]	% Difference
ErbisicR_H =	0.84	12.47	6.85	-5.62	45.10
ErbisicR_H ⊥	0.84	10.67	5.30	-5.37	50.34
ErbisicR_L =	0.89	7.30	6.16	-1.14	15.66
ErbisicR_L ⊥	0.89	6.98	4.07	-2.91	41.69

Both methodologies confirm (Figure 11) an anisotropy in foam thermal conductivity. Difference between longitudinal and transverse K_{eff} is higher for ErbisicR_H than for ErbisicR_L.
According to [15.][16.], measurement based on the elongated tetrakaidecahedron cell in Figure 7 have been performed on 10 different cells chosen randomly on each foam and averaged.

Table 5 Measured tetrakaidecahedron cell parameters

Specimen	D [mm]	H [mm]	l [mm]	Strut thickness [mm]	Aspect ratio H/D
ErbisicR_H	5.08	7.48	2.61	1.06	1.48
ErbisicR_L	4.69	6.60	2.24	0.72	1.41

ErbisicR_H aspect ratio is higher than ErbisicR_L meaning a more elongated cell shape, that could qualitatively explain the more anisotropic behaviour of ErbisicR_H foam. Difference between longitudinal and transverse K_{eff} is higher for experimental then for simulated data. A possible explanation is that meshing procedure from acquisition to final computational model, was done with several smoothing steps.

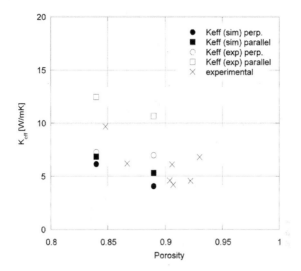

Figure 11 Experimental and simulated K_{eff} values

Simulation results evidence the foams anisotropic behaviour. This anisotropy is even more evident in the experimental values.

CONCLUSIONS

Effective thermal conductivity FE simulations were performed on Si-SIC foams with computational domains reconstructed via X ray Computed Tomography.
K_{eff} values from FE were than compared with experimentally measured data on the very same foams. Data show a marked anisotropy in thermal conductivity.

Discrepancies between numerical and experimental thermal conductivity values were also found. On the experimental side further work will be devoted in measuring the bulk thermal conductivity of the foam material, this will be done by measuring it, by laser flash methods, on Si-SiC disks. On the numerical side further calibration work is ongoing to reduce morphological differences between the real foams and the acquired-meshed ones. Further work is also ongoing trying to discern into the strut, and thus efficiently segment, Si-SiC ceramic and metallic Si, having both similar attenuation coefficient.

ACKNOWLEDGMENTS

Authors wish to thank Erbicol SA for providing ceramic foams for the experiments. Special thanks goes also to our colleagues Maurizio Barbato, for the indispensable suggestions on experimental set up design and data interpretation, and Luca Diviani for his advise on meshing and FE analysis.

BIBLIOGRAPHY

[1.]   Ashby, M. F., A. Evans, N. A. Fleck, L. J. Gibson, J. W. Hutchinson, and H. N. G.
        Wadley, *Metal Foams - A Design Guide*, Butterworth-Heinemann (2000).
[2.]   Bastawros, A. F., "Effectiveness of Open-Cell Metallic Foams for High Power Electronic
        Cooling," *ASME Heat Transf. Division*, **361**, 211 (1998a).
[3.]   C.Y. Zhao, W. Lu, S.A. Tassou, Int. J. Heat Mass Transfer 49,2762 (2006)
[4.]   E.A. Moreira, M.D.M. Innocentini, J.R. Coury, J. Eur. Ceram. Soc. 24, 3209 (2004)
[5.]   P. G. Collishaw, J. R. G. Evans"An assessment of expressions for the apparent thermal
        conductivity of cellular materials" J. Mat. Science, 29 (1994) 486-498
[6.]   J.W. Peak, B.H. Kang, S.Y. Kim, J.M. Hyun, Int. J. Thermophys. 21(2), 453 (2000)
[7.]   K.Boomsma, D.Poulikakos, Y.Ventikos, simulation of flow through open cell metal foam
        using an idealized periodic cell structure, Int. J. of Heat and fluid Flow 24, (2003) 825-
        834
[8.]   Radon, J., Über die Bestimmung von Funktionen durch ihre integralwerte längs gewisser
        Mannigfaltigkeiten. Berichte über die Verhandlungen der Königlich Sächsischen
        Gesellschaft der Wissenschaften zu Leipzig, Math. Phys. Klasse, volume 69, (1917), pp.
        262–277.
[9.]   E.Boutsianis et al, Computational simulation of intracoronary flow based on real coronary
        geometry, European Journal of Cardio-thoracis Surgery 26, (2004), 248-256
[10.]   Schmidt J, Scheiffele M, Mach A, von Issendorf F, Ortona A, Manufacturing,
        characterization and testing of Si-Sic foams for porous burner applications
        (PCM2005),2005
[11.]   Radon, J., Über die Bestimmung von Funktionen durch ihre integralwerte längs gewisser
        Mannigfaltigkeiten. Berichte über die Verhandlungen der Königlich Sächsischen
        Gesellschaft der Wissenschaften zu Leipzig, Math. Phys. Klasse, volume 69, (1917), pp.
        262–277.
[12.]   K. Schwartzwalder, A. V. Somers: Method of making porous ceramic articles, US patent
        No. 3090094, 1963
[13.]   Natterer, F., The mathematics of computerized tomography, John Wiley & Sons Inc, New
        York, 1986.
[14.]   Kak, A., Slaney, M., Principles of Computerized Tomographic Imaging, IEEE Press, New
        York, 1987.
[15.]   R.M. Sullivan, L.J. Ghosn, B.A. Lerch, "A general tetrakaidecahedron model for open-
        celled foams" INT. J. of solids and structures, 45 (6): 1754-1765 Mar 15 2008
[16.]   R.M. Sullivan, L.J. Ghosn, B.A. Lerch, "Elongated Tetrakaidecahedron Micromechanics
        Model for Space Shuttle External Tank Foams", NASA/TP—2009-215137
[17.]   V.V.Calmidi, R.L. Mahajan, The effective thermal conductivity of High Porosity Fibrous
        Metal Foams, Transaction of the ASME, vol.121, 1999
[18.]   W. Xu, H. Zhang, Z. Yang, J. Zhang, "The effective Thermal conductivity of three
        dimensional reticulated foam materials", J Porous Mater (2009) 16:65-71
[19.]   M. Scheffler and P. Colombo eds. "Cellular Ceramics: Structure, Manufacturing,
        Properties and Applications", WILEY-VCH Verlag GmbH, Weinheim, Germany, 2005

THE EFFECTS OF β-Si$_3$N$_4$ SEEDING AND α-Si$_3$N$_4$ POWDER SIZE ON THE DEVELOPMENT OF POROUS β-Si$_3$N$_4$ CERAMICS

Daniel A. Gould and Kevin P. Plucknett[*]
Dalhousie University, Materials Engineering Program, Department of Process Engineering and Applied Sciences, 1366 Barrington Street, Halifax, Nova Scotia, B3J 2X4, Canada.

Liliana B. Garrido
CONICET, Centro de Technologia de Recursos Minerales y Cerámica (CETMIC, CIC-CONICET-UNLP), Cam. Centenario y 506, C.C.49 (B 1897 ZCA) M.B. Gonnet. Pcia. De Buenos Aires, ARGENTINA

Luis A. Genova
IPEN Instituto de Pesquisas Energéticas e Nucleares, CCTM Centro de Ciência e Technologia de Materiais, Cidade Universitária, Travessa R 400, 05508-900 São Paulo, BRAZIL

ABSTRACT

Porous silicon nitride (Si$_3$N$_4$) ceramics have significant potential in a number of engineering applications due to their potential for favorable mechanical properties, even with moderately high porosities. In the present work two methods of microstructural control were assessed; (i) seeding with β-Si$_3$N$_4$ grains and (ii) varying the α-Si$_3$N$_4$ starting powder size. The aim of this work was to develop a ceramic with an improved microstructure, resulting in increased strength. The β-Si$_3$N$_4$ seed grains were prepared 'in-house' using a simple, moderately low-temperature processing approach. A series of nominally bi-modal α-Si$_3$N$_4$ powder mixtures were also examined, using various ratios of powders with particle sizes of approximately 0.3 and 1.0 μm. It was demonstrated that the β-Si$_3$N$_4$ aspect ratio was predominantly influenced by the amount of coarse α-Si$_3$N$_4$ powder in the starting mixture, rather than the amount of β-Si$_3$N$_4$ seed grains. The highest aspect ratios (up to ~12:1) were obtained when using 100 % coarse α-Si$_3$N$_4$, as the α- to β-Si$_3$N$_4$ transformation period is prolonged. The strength was also slightly increased with a higher coarse α-Si$_3$N$_4$ fraction in the starting mixture. In this instance, seeding was shown to provide a small benefit, with the highest strengths (up to ~350 MPa) observed for samples prepared with 0.5 and 1 wt. % β-Si$_3$N$_4$ seed crystals.

INTRODUCTION

Silicon nitride (Si$_3$N$_4$) ceramics are an important family of engineering ceramics that are used in a wide variety of applications, including bearings and seals, turbocharger rotors, etc.[1] While most of these uses are based on dense β-Si$_3$N$_4$, there is growing interest in the development and characterization of porous β-Si$_3$N$_4$ ceramics. Porous β-Si$_3$N$_4$ can be produced using a variety of approaches, with the ultimate aim of generating either micro- or macro-porosity. The generation of macro-porosity can be achieved through the use of pyrolyzable filler particles, which can be readily removed via heat-treatment prior to sintering.[2] One of the simplest methods to generate micro-porosity involves partial sintering. In this instance sintering additive selection is made in such a way that densification is largely retarded, while the α- to β-Si$_3$N$_4$ phase transformation is still able to proceed. This can be achieved through the use of single, refractory oxide sintering aid (e.g. single rare earth oxide additions).[3-7] Using such an approach, high performance, porous β-Si$_3$N$_4$ ceramics have been developed, with strengths up to ~450 MPa for samples prepared with 35 to 50 % porosity.[3] More

---

[*] Corresponding Author (email: kevin.plucknett@dal.ca)

recently it has been demonstrated that the use of a low volume fraction of multiple sintering additives can achieve similar results while lowering the processing temperature.[8,9]

It has been shown when using a low volume fraction of multiple sintering additives, based on rare earth oxides mixed with magnesia, that the choice of rare earth oxide can have a significant effect on the final β-Si₃N₄ grain aspect ratio.[9] In particular, the highest β-Si₃N₄ aspect ratios were observed when using La₂O₃ additions, and the aspect ratios were found to increase with increasing rare earth ionic radius in the sequence Yb < Y < Nd < La. Comparable results have also been observed for dense ceramics by a number of authors.[10,11] At the same time, while the α- to β-Si₃N₄ phase transformation is strongly influenced by the rare earth oxide used for single additives, when combined with MgO in porous β-Si₃N₄ this effect is largely negated and the transformation behavior is nominally identical, regardless of the rare earth used.[9] Consequently, the differences in grain aspect ratio development can be attributed solely to the rare earth selected, and the influence of the rare earth on Si and N attachment behavior for both the prism and basal planes, in accordance with dense β-Si₃N₄ ceramics.[12]

In the present work, the multiple sintering aid approach has again been taken, utilizing a mixture of magnesia (MgO) with either yttria (Y₂O₃) or lanthana (La₂O₃). In order to increase the potential degree of flexibility in microstructure control, two additional processing modifications have been taken. For select samples, β-Si₃N₄ seeds have been added, with the seeds produced 'in-house' using a moderately low-temperature synthesis step that allows relatively easy manufacture of the seeds.[13] The seeding approach has been widely employed in the production of dense β-Si₃N₄ ceramics, starting with the early work of Hirao and colleagues,[14,15] and more recently for the preparation of anisotropic porous β-Si₃N₄.[16] In addition, the influence of the α-Si₃N₄ starting powder particle size has also been investigated in the current study, using powders with nominal mean particle sizes of ~0.3 and/or ~1.0 μm, blended in a variety of ratios. This approach is designed to slow the α- to β-Si₃N₄ transformation rate, thereby altering the microstructural development. The microstructure and grain aspect ratio of the sintered samples were assessed in combination with the fracture strength, which has been measured on 'as-sintered' samples to ascertain the influence of the as-prepared surface condition; it is anticipated that such conditions would be employed in low-cost applications of these materials.

EXPERIMENTAL PROCEDURES
Raw Materials

Samples have been prepared using both Ube SN E-10 and SN E-03 α-Si₃N₄ powders (Ube Industries, New York, NY), with nominal particle sizes of 0.3 and 1.0 respectively. Sintering additives of Y₂O₃ and La₂O₃ were obtained from Metall Rare Earth (Shenzhen, China), and had a nominal purity of 99.99 at. %, while MgO was obtained from Inframat Advanced Materials (Farmington, CT). In addition, β-Si₃N₄ seed crystals were prepared using the simple method outlined in Gould et al.[13] Briefly, this approach utilizes multiple oxide additives (1.5 Y₂O₃, 0.25 MgO and 0.25 CaO (wt. %)) to produce seeds at moderately low temperatures (i.e. 1625°C for 4 hours), allowing the seed grains to be cleaned and separated using a dilute, room temperature rinse with hydrofluoric (HF) acid. The seed crystals, which were fully transformed to β-Si₃N₄ under these conditions, had an average aspect ratio of ~4:1 and a typical length of 3-4 μm.[13]

Milling and Sintering Procedures

Blends of the two α-Si₃N₄ powders have been prepared in the ratios (fine:coarse) 100:0, 75:25, 50:50, 25:75 and 0:100 by ball milling in iso-propyl alcohol for 24 h using tetragonal zirconia milling media. In each case sintering additives of 2.5 wt. % Y₂O₃ and 0.5 wt. % MgO (La₂O₃ was substituted for Y₂O₃ on a molar equivalent basis) were used. In addition to 'seed-free' compositions, samples were also prepared with 0.5, 1, 2 or 5 wt. % of β-Si₃N₄ seed crystals (with an equivalent reduction in the total α-Si₃N₄ content). A summary of the samples that were prepared (for *both* Y₂O₃ and La₂O₃ systems) is provided in Table 1. After milling, the powders were dried and sieved to -75 μm. Pellets with a diameter of ~31.75 mm were then uniaxially pressed (~31 MPa) from the sieved powder and these were subsequently cold-isostatically pressed (~170 MPa). For sintering, samples were sited in a 50 α-Si₃N₄/49BN/1MgO (wt. %) powder bed, within a graphite crucible. Sintering was conducted in a static nitrogen atmosphere (0.1 MPa), at a temperature of 1650°C, and with a final hold period at temperature of 2 hours.

Table 1. A summary of the Si₃N₄ samples prepared in the present work, for both the Y₂O₃ and La₂O₃ based systems, as a function of β-Si₃N₄ seed content and the ratio of fine:coarse α-Si₃N₄ powder (Ube SN E-10 and E-03, respectively). Note that compositions in the shaded area were not examined in the present work.

Seed content (wt. %)	100:0	75:25	50:50	25:75	0:100
0	✓	✓	✓	✓	✓
0.5	✓				✓
1	✓				✓
2	✓	✓	✓	✓	✓
5	✓				✓

Characterization of Porous β-Si₃N₄ Ceramics

The densities of sintered samples were determined by mercury immersion. Microstructural characterization was performed using field emission scanning electron microscopy (SEM; Hitachi S-4700, Hitachi High Technologies, Inc., Tokyo, Japan), on fracture surfaces coated with a thin layer of evaporated carbon. X-ray diffraction (XRD; Bruker D-8 Advance,) was used to assess the α- to β-Si₃N₄ phase transformation. The β-Si₃N₄ grain aspect ratio was determined using SEM on grains liberated from the sintered samples using a dilute HF acid soak. The strength of the as-sintered samples (i.e. no post-sinter surface machining) was measured using the ball-on-ring biaxial flexure test, following ASTM C-1499. An average of five or six biaxial samples was tested for each of the compositional and processing variants examined.

RESULTS AND DISUSSION

Microstructure of Sintered Samples

All of the samples that were sintered at 1650°C, outlined in Table 1, exhibited complete α- to β-Si₃N₄ transformation. The extent of porosity varied from approximately 18 to 30 vol. %, depending on the processing variables. The porosity level was observed to increase with either increasing β-Si₃N₄ seed content or when the α-Si₃N₄ starting powder mixture was coarser in nature (i.e. a higher SN E-03 fraction). Conversely, a higher fraction of the finer SN E-10 α-Si₃N₄ powder in the starting mixture resulted in higher sintered densities. The effects of selecting either the Y₂O₃ or La₂O₃ based additive systems had a negligible influence on the sintered densities in comparison to the other variables. The effects of starting powder upon the final microstructure are highlighted for the Y₂O₃-MgO additive samples in Figure 1. The use of 100 % SN E-03 (Figure 1(b))), the coarser of the two powders, results in a generally more coarse final microstructure than when the finer Ube SN E-10 starting powder is employed (Figure 1(a)). For the case of the coarser starting powder, it can be expected that there will be less 'pre-formed' β-Si₃N₄ crystals present in the original powder, such that there are effectively less potential 'nuclei' for β-Si₃N₄ growth. This is coupled with the fact that the larger particles can be expected to take a longer time to dissolve, due to their lower surface area to volume ratio than the finer powder. In combination, this leads to a more coarse final structure when the SN E-03 powder is predominantly used. Qualitatively, it also appears that the aspect ratio is slightly greater in the case of the coarse α-Si$_3$N$_4$ powder. This may be anticipated through the potential for reduced growth hindrance when there are less initial 'nuclei', as there will initially be a lower concentration of β-Si$_3$N$_4$ grains locally for direct contact. In addition, the actual α- to β-Si$_3$N$_4$ phase transformation process is expected to be slowed as the larger α-Si$_3$N$_4$ particles will take longer to dissolve. Generally similar microstructural observations were also made for the La$_2$O$_3$-MgO system.

In comparing the effects of using either Y$_2$O$_3$ or La$_2$O$_3$, both in combination with MgO and using the same β-Si$_3$N$_4$ starting powder, it is clear that there is little difference in terms of final grain size (Figure 2). Qualitatively, it appears that the La$_2$O$_3$-based system (Figure 2(a) exhibits a slightly greater grain aspect ratio than for the Y$_2$O$_3$-based example (Figure 2(b)), which has been previously noted for both porous and dense β-Si$_3$N$_4$ ceramics.[9-11] The higher aspect ratios observed in La$_2$O$_3$-based systems have been attributed to the change in atomic attachment behavior of the rare earth element, which hinders Si and N attachment to the prism planes, thereby promoting elongated growth parallel to the c-axis direction.[12]

(a)   (b)

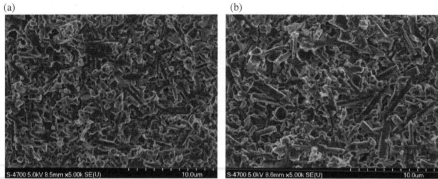

Figure 1. SEM images of samples prepared with Y$_2$O$_3$-MgO sintering addition using (no β-Si$_3$N$_4$ seed grains): (a) 100 % Ube SN E-10 starting powder, and (b) 100 % SN E-03 starting powder.

Quantitative assessment of the β-Si$_3$N$_4$ grain dimensions is required to more fully discuss the microstructure evolution as a function of the examined processing variables. Figure 3 highlights the effects of both additive type and the ratio of coarse:fine α-Si$_3$N$_4$ powder on the measured β-Si$_3$N$_4$ grain aspect ratio. It is clear that increasing the coarse fraction of α-Si$_3$N$_4$ powder in the starting mixture increases the aspect ratio. Similarly, the selection of La$_2$O$_3$ rather than Y$_2$O$_3$ also increases the aspect ratio, although the effect is lesser than noted in a prior publication for samples processed at 1700°C.[9]

Mechanical Property Evaluation

In the present work the method of biaxial flexure testing has been selected for the primary aspect of ease of sample preparation. In addition, it was planned to assess all samples in their 'as-sintered' state (i.e. no post-sinter surface machining). The primary reason behind this approach was to minimize potential processing costs in terms of producing a low-cost, high performance filter material. With respect to materials processing variants examined, all of the examples outlined in Table 1 have been tested, for both Y$_2$O$_3$-MgO and La$_2$O$_3$-MgO systems. In particular, two trends have been examined, the influence of starting powder size ratio and the effects of β-Si$_3$N$_4$ seeds.

For the case of La$_2$O$_3$-MgO additions, the effects of the fine:coarse powder size ratio is shown in Figure 4(a), for both 0 and 2 wt. % β-Si$_3$N$_4$ seeds. It is clear that there is a moderately small influence of the starting powder blend on the strength of the samples, with an increasing coarse α-Si₃N₄ fraction giving rise to increased strength (even though these samples possess lower sintered densities). This observation is matched with the increasing grain aspect ratios, shown in Figure 3(a), although the effects are more subtle for the biaxial flexure strength. This general trend is consistent with the fact that increasing aspect ratio can be seen to increase the extent of mechanical interlocking of the grains, which results in small increases in strength. Interestingly, adding 2 wt. % seed does not result in any significant improvements in mechanical behavior, while the highest seed contents (i.e. 5 wt. %) can actually be slightly detrimental.

(a)   (b)

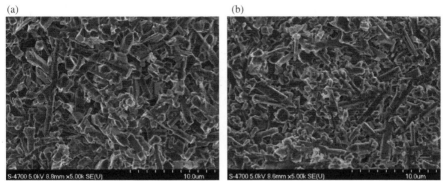

Figure 2. SEM images of samples prepared with 100 % Ube SN E-10 starting powder (no β-Si₃N₄ seed grains) sintering using sintering additions of: (a) La₂O₃-MgO and (b) Y₂O₃-MgO.

(a)   (b)

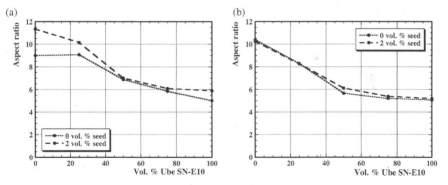

Figure 3. β-Si₃N₄ grain aspect ratios for samples prepared with various fine:coarse α-Si₃N₄ powder ratios and either 0 or 2 wt.% β-Si₃N₄ seeds, prepared using sintering additions of: (a) La₂O₃-MgO and (b) Y₂O₃-MgO.

The effects of adding β-Si₃N₄ seeds to samples prepared with either 100 % fine or 100 % coarse α-Si₃N₄ powder are shown in Figure 4(b). It is apparent that the addition of high β-Si₃N₄ seed contents (i.e. 5 vol. %) does not result in any benefits in terms of biaxial flexure strength. The most significant benefits are for samples with 0.5 to 2 wt. % of β-Si₃N₄ seeds when using the coarse α-Si₃N₄ powder, and 0.5 or 1 wt. % seeds for the fine.

In comparison to the situation with La$_2$O$_3$-MgO additions, the Y$_2$O$_3$-MgO additive samples demonstrated generally lower strengths. Figure 5(a) highlights the influence of the fine:coarse α-Si$_3$N$_4$ powder ratio on the biaxial strength for Y$_2$O$_3$-MgO additions. For seeded examples, the strength is largely unaffected by the ratio of fine:coarse starting powder. Conversely, for the unseeded samples there is a clear maximum in observed strength for the 50:50 ratio mixture of powders. The reason for this is not clear at the present time, although it may be related to marginal warping of the disc-shaped samples; while the vast majority of samples showed no warping, a limited number were observed to show a minimal amount of distortion during sintering. It is likely that warping can be avoided through the careful use of a pre-packed powder bed base, on to which the samples would be placed.

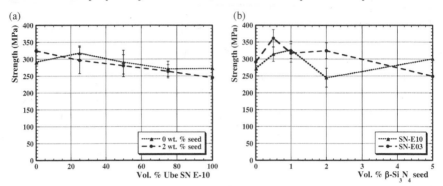

Figure 4. (a) The effects of fine:coarse α-Si$_4$N$_4$ powder content on the biaxial flexure strength of samples prepared with La$_2$O$_3$-MgO sintering additions. (b) The effects of β-Si$_3$N$_4$ seed content on the biaxial flexure strength of samples prepared with La$_2$O$_3$-MgO sintering additions and either 100% fine or 100 % coarse α-Si$_3$N$_4$ starting powder.

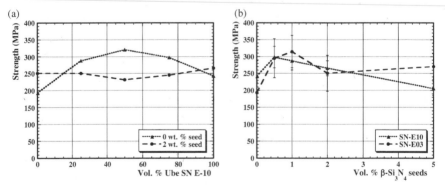

Figure 5. (a) The effects of fine:coarse α-Si$_4$N$_4$ powder content on the biaxial flexure strength of samples prepared with Y$_2$O$_3$-MgO sintering additions. (b) The effects of β-Si$_3$N$_4$ seed content on the biaxial flexure strength of samples prepared with Y$_2$O$_3$-MgO sintering additions and either 100% fine or 100 % coarse α-Si$_3$N$_4$ starting powder.

CONCLUSIONS

In the present work, a range of porous β-Si$_3$N$_4$ ceramics has been developed, with an emphasis on modifying the microstructure through relatively simple approaches. In an effort to achieve this, microstructural control methods have been investigated through both the incorporation of β-Si$_3$N$_4$ seeds, manufactured 'in-house', and by changing the rate at which the starting powder can dissolve into the oxynitride glass that is present at the sintering temperature, through the use of fine or coarse α-Si$_3$N$_4$ starting powders, or a mixture of the two. Pressureless sintering at 1650°C for a period of 2 hours resulted in samples that were completely transformed to β-Si$_3$N$_4$ and exhibited a retained porosity of up to 30 vol. %, depending on processing variables. Generally, samples prepared with La$_2$O$_3$-MgO additions exhibited a slightly higher β-Si$_3$N$_4$ grain aspect ratio than those prepared with Y$_2$O$_3$-MgO additions. In terms of the starting powder mixtures, higher aspect ratios were also achieved when predominantly coarse α-Si$_3$N$_4$ powders are used (i.e. SN E-03). A similar response was also noted with the biaxial strength, with samples prepared from more coarse powder mixtures exhibiting higher strengths, even though they also exhibited lower sintered densities. The effects of β-Si$_3$N$_4$ seeding were somewhat more subtle, with a peak in strength typically observed when between 0.5 and 1 wt. % was employed. Higher seed contents resulted in strengths comparable to the two unseeded baseline materials (either 100 % fine or 100 % coarse), and even slight strength degradation relative to the unseeded samples.

ACKNOWLEDGEMENTS

The Natural Sciences and Engineering Research Council of Canada (NSERC) are gratefully acknowledged for funding this work through both the Inter-Americas Collaborative Research in Materials (CIAM) and Undergraduate Student Research Awards (USRA) programs. The support of the Canada Foundation for Innovation, the Atlantic Innovation Fund, and other partners who helped fund the Facilities for Materials Characterisation, managed by the Dalhousie University Institute for Materials Research, is also gratefully acknowledged.

REFERENCES

[1]A. Okada, 'Automotive and Industrial Applications of Structural Ceramics in Japan,' *J. Eur. Ceram. Soc.*, **28** [5] 1097 (2008).

[2]A. Diaz and S. Hampshire, 'Characterisation of Porous Silicon Nitride Materials Produced with Starch,' *J. Eur. Ceram. Soc.*, **24** [2] 413 (2004).

[3]C. Kawai and A. Yamakawa, 'Effect of Porosity and Microstructure on the Strength of Si$_3$N$_4$: Designed Microstructure for High Strength, High Thermal Shock Resistance, and Facile Machining,' *J. Am. Ceram. Soc.*, **80** [10] 2705 (1997).

[4]K.P. Plucknett and M.H. Lewis, 'Toughness Enhancement of Green-State Silicon Nitride by Pre-Sinter Heat-Treatment,' *J. Mater. Sci. Lett.*, **17** [23] 1987 (1998).

[5]J.H. She, J.-F. Yang, D.D. Jayaseelan, N. Kondo, T. Ohji, S. Kanzaki and Y. Inagaki, 'Thermal Shock Behavior of Isotropic and Anisotropic Porous Silicon Nitride,' *J. Am. Ceram. Soc.*, **86** [4] 738 (2003).

[6]Z.Y. Deng, Y. Inagaki, J.H. She, Y. Tanaka, Y.F. Liu, M. Sakamoto and T. Ohji, 'Long Crack R-curve of Aligned Porous Silicon Nitride,' *J. Am. Ceram. Soc.*, **88** [2] 462 (2005).

[7]J. Yang, J.-F. Yang, S.-Y. Shan, J.-Q. Gao and T. Ohji 'Effect of Sintering Additives on Microstructure and Mechanical Properties of Porous Silicon Nitride Ceramics,' *J. Am. Ceram. Soc.*, **89** [12] 3843 (2006).

[8]M. Quinlan, L.B. Garrido, L.A. Genova and K.P. Plucknett, 'Compositional Design of Porous β-Si$_3$N$_4$ Prepared by Pressureless-Sintering Compositions in the Si-Y-Mg-(Ca)-O-N System,' *Ceram. Eng. Sci. Proc.*, **28** [9] 49 (2007).

[9]K.P. Plucknett, M. Quinlan, L.B. Garrido and L.A. Genova, 'Microstructural Development in Porous β-Si$_3$N$_4$ Ceramics Prepared with Low Volume RE$_2$O$_3$-MgO-(CaO) Additions (RE = La, Nd, Y, Yb),' *Mater. Sci. Eng. A*, **489** [1-2] 337 (2008).

[10]R.L. Satet, M.J. Hoffmann and R.M. Cannon, 'Experimental Evidence of the Impact of Rare-Earth Elements on Particle Growth and Mechanical Behaviour of Silicon Nitride,' *Mater. Sci. Eng. A*, **422** [1-2] 66 (2006).

[11]P.F. Becher, G.S. Painter, N. Shibata, S.B. Waters and H.-T. Lin, 'Effects of Rare-Earth (RE) Intergranular Adsorption on the Phase Transformation, Microstructure Evolution, and Mechanical Properties in Silicon Nitride with RE$_2$O$_3$ + MgO Additives: RE=La, Gd, and Lu,' *J. Am. Ceram. Soc.*, **91** [7] 2328 (2008).

[12]G.S. Painter, F.W. Averill, P.F. Becher, N. Shibata, K. van Benthem and S.J. Pennycook, 'First-Principles Study of Rare Earth Adsorption at β-Si$_3$N$_4$ Interfaces,' *Phys. Rev. B*, **78** [21] 4206 (2008).

[13]D.A. Gould, M. Quinlan, M.P. Albano, L.B. Garrido, L.A. Genova and K.P. Plucknett, 'Low-Temperature Synthesis of Acicular β-Si$_3$N$_4$ Seed Crystals,' *Ceram. Int.*, **35** [4] 1357 (2009).

[14]K. Hirao, T. Nagaoka, M.E. Brito and S. Kanzaki, 'Microstructure Control of Silicon Nitride by Seeding with Rod-like β-Silicon Nitride Particles,' *J. Am. Ceram. Soc.*, **77** [7] 1857 (1994).

[15]K. Hirao, M. Ohashi, M.E. Brito and S. Kanzaki, 'Processing Strategy for Producing Highly Anisotropic Silicon Nitride,' *J. Am. Ceram. Soc.*, **78** [6] 1687 (1995).

[16]Y. Inagaki, Y. Shigegaki, M. Ando and T. Ohji, 'Synthesis and Evaluation of Anisotropic Porous Silicon Nitride,' *J. Eur. Ceram. Soc.*, **24** [2] 197 (2004).

DEVELOPMENT OF NOVEL MICROPOROUS ZrO_2 MEMBRANES FOR H_2/CO_2 SEPARATION

Tim Van Gestel, Felix Hauler, Doris Sebold, Wilhelm A. Meulenberg, Hans-Peter Buchkremer
Forschungszentrum Jülich GmbH, Institute of Energy Research, IEF-1: Materials Synthesis and Processing, Leo-Brandt-Strasse, D-52425 Jülich, Germany

ABSTRACT
This paper gives an overview of diverse microporous ceramic membranes which are under investigation at IEF-1 for H_2/CO_2 separation in power plants. A summary is given of what has been accomplished in the synthesis of multilayer membranes in our laboratory and the progress in the area of thin-film deposition of silica and non-silica membranes is illustrated. These membranes have to combine the required gas separation factor and permeation and contaminants such as water and acidic vapors must be dealt with. Stabilized microporous SiO_2 membranes are one of the candidate materials due to their demonstrated resistance to water vapor. In our lab, we achieved a high H_2/CO_2 selectivity for such membranes by optimizing all conditions in the manufacturing procedure (support, sol preparation, mesoporous intermediate layers, cleanroom coating). In addition, the development of an Y_2O_3-stabilized ZrO_2 membrane for application under harsher conditions, including a high amount of acid contaminents, is also reported.

INTRODUCTION
1.Power plant concept - H_2/CO_2 Separation

A considerable amount of research activities are focussed worldwide on the development of technologies to reduce the emission of CO_2 in the atmosphere. This reduction could be achieved by introducing gas separation membranes in fossil power plants, which is according to a number of sources associated with lower efficiency losses compared with conventional separation technologies. Although membranes are already used for a wide range of separation applications in other fields (e.g. water purification, chemical industry), membranes for separating the technically relevant gases in fossil power plants are still far from being suitable for implementation in industrial applications.

A concept which has a good potential for industrial implementation is the precombustion concept. The precombustion concept involves removing all or part of the carbon content of a fuel before burning it. The first step of precombustion is to convert the fuel into a synthesis gas that primarily consists of CO and water. Next, the carbon monoxide is reacted with steam in a shift reactor to produce CO_2 and hydrogen. The CO_2 is then separated and transported to a storage site and the remaining hydrogen-rich gas is combusted in a gas turbine to generate electricity, which results in a flue gas that consists of water.

An advantage of the precombustion system is that it could be optimized for generating electricity, hydrogen and chemicals in one plant. The conventional method for CO_2 separation is to use a solvent adsorption, but this method leads to large energy and environmental penalties. The desired properties of a membrane for this application would include the ability to withstand operating temperatures in excess of 200°C, high operating pressures and components of the shifted feed gas stream, e.g. CO_2, H_2, H_2O, H_2S. Further, the membranes must combine a high H_2 permeability with a sufficient H_2/CO_2 selectivity at temperatures in excess of 200°C.

2. Development of microporous membranes

Microporous membranes are considered by many research groups as one of the candidates for separation problems involving small gas molecules (He, H$_2$, N$_2$, CO, CO$_2$, CH$_4$, …), because they hold the potential to combine an excellent selectivity with a high gas flow. Typically, microporous membranes are developed as an alternative for polymeric gas separation membranes, for industrial applications in which polymeric membranes can not perform well or do not have the required lifetime. Examples are applications including acid contaminents, steam, high temperatures and high pressures. Polymeric membranes (e.g. polyimide, polysulfon) are generally considered for another power plant concept (post-combustion) which involves the separation of CO$_2$ and N$_2$ [1]. The operation temperature for the membrane devices in the precombustion concept is in the range 200 – 350°C and therefore polymeric membranes are not considered for this application.

The main classes of microporous membranes include amorphous SiO$_2$ and doped SiO$_2$ membranes, templated SiO$_2$ membranes and zeolite membranes. In this research field (H$_2$/CO$_2$ separation), amorphous SiO$_2$ membranes have been the most extensively investigated. Amorphous SiO$_2$ membranes are basically synthesized through two methods: sol-gel coating and chemical vapor deposition (CVD) [2,3]. From a practical point of view, sol-gel coating appears as a very interesting method, since the sol coating methods are the same as conventional powder suspension coating methods and exhibit the same advantages (inexpensive in terms of capital costs, simplicity of the equipment). There are a variety of coating methods used for making sol-gel membranes among which spin-coating (flat membranes) and dip-coating (tubular membranes) are the most frequently used. CVD membranes have frequently a higher selectivity for H$_2$, but these membranes have also usually a lower permeability. It should however be noticed that literature data on permeability of SiO$_2$ membranes show a large scattering, making a comparison for different types of membranes very difficult. A problem which is often reported for sol-gel coating is a lack of reproducability, but this problem is probably often due to inaccurate working conditions and also the lack of clean room conditions.

An important drawback frequently mentioned for amorphous SiO$_2$ membranes is however a degradation of the membrane material in water containing atmospheres. For this reason, a number of researchers focused on developing alternative SiO$_2$ materials with an improved stability. One common approach is to improve the material stability by adding doping compounds, such as ZrO$_2$, TiO$_2$, Al$_2$O$_3$, NiO, etc. An extension of the doped silica membrane work is the synthesis of metal doped membranes (e.g. Ref. 4 - 8). Another common method which could be potentially used to improve the material stability is the synthesis of hybrid organic-ceramic SiO$_2$ membranes (e.g. Ref. 9 - 11). Another drawback could be the limited thermal stability at higher temperatures of a number of membranes reported in literature. Frequently, these membranes are fired by a very low thermal treatment – sometimes just 300°C – and/or in an inert atmosphere, which implies that the membrane can be sensitive towards structural changes when afterwards applied at higher temperatures or in an oxidizing atmosphere.

Another interesting approach reported by many researchers is to prepare crystalline microporous SiO$_2$ based membranes (zeolites) (e.g. Ref. 12,13). Some of these membranes exhibit fascinating properties including a well defined pore structure with a pore size of approximately 0.5 nm and are not susceptible to densification at higher temperatures or degradation in steam. Several research groups succeeded to prepare zeolite membranes with a good performance in gas and liquid separation, but the existence of intercrystalline pores is often reported as a factor that limits the separation efficiency for small gas molecules. On the other hand, a number of membranes have been described which give a significant separation for CO$_2$, especially when the measuring temperature approximates room temperature. This result is explained based on the high CO$_2$ sorption capacity of the zeolitic membrane material and the occurence of a surface diffusion mechanism.

While many of the research efforts are associated with Si-based membranes, it may eventually prove necessary to search for alternative materials with a higher stability. Due to the high chemical resistance of some non-silica materials in diverse membrane applications (e.g. TiO$_2$ and ZrO$_2$ membranes in ultra- and nano-filtration) research efforts have been devoted in our lab the past years to the potential application of non-silica membranes as an alternative for amorphous silica membranes in gas separation. Contrary to silica membranes, non-silica membranes with a connected network of micropores and a selectivity for small gases are not reported yet. In membrane filtration applications, zirconia is generally recognized for its chemical stability as a membrane material and with the addition of a dopant (e.g. Y) also a high thermal stability is achieved. Therefore, we have chosen to develop a membrane consisting of stabilized zirconia (8Y$_2$O$_3$-ZrO$_2$). The approach has been to develop methods for producing 8Y$_2$O$_3$-ZrO$_2$ membrane layers, with the same properties as microporous SiO$_2$ toplayers. In order to understand the structure and the gas transport mechanisms of these membranes, more conventional SiO$_2$ membranes are always prepared in parallel on the same membrane supports and tested.

An extension of the H$_2$ separation membrane work in our laboratory is a study on the synthesis and potential application of modified silica membranes. From contact with energy companies, we learned that several stages come into focus for a membrane separation unit in the power plant process and that the operating conditions in these stages are very different. Therefore, the following strategy is adopted. First, silica and modified silica membranes are studied for application at low steam pressures. Second, novel non-silica membranes are studied for application at high steam pressures.

EXPERIMENTAL

The development of all multilayer membranes in our work involves basically three fields of research including: (1) developing a high-quality support material, (2) developing mesoporous intermediate layers, and (3) developing a functional microporous toplayer.

1. Macroporous support

The manufacturing procedure of the support is very simple and includes: (1) vacuum-casting of a powder suspension into disks with a thickness of ~ 3 mm and a diameter of ~ 50 mm (Sumitomo α-Al$_2$O$_3$ or Tosoh 8Y$_2$O$_3$-ZrO$_2$ powder), (2) sintering the green disks at 1100°C, (3) grinding, and (4) polishing one side with diamond paste (Struers, DP-Paste 6 μm and 3 μm). The final diameter and thickness of the obtained porous support disk measure ~ 40 mm and ~ 26 mm, respectively. Post-treatment includes ultra-sonic washing in acetone and firing at 900°C for 1h.

2. Sol synthesis

Mesoporous and microporous membrane layers were prepared through two different methods: (1) a colloidal particle sol method and (2) a polymeric sol method.

Mesoporous membrane layers were made by a colloidal sol-gel coating procedure. In a first coating experiment, alumina membrane layers were prepared from sols containing γ-alumina colloidal particles with a size of ~ 30 nm. The sol preparation was based on the well-known Yoldas process, which includes hydrolysis of a metal-organic precursor (Al(OC$_4$H$_9$)$_3$, Sigma-Aldrich) with H$_2$O and subsequent destruction of larger agglomerates with HNO$_3$ at elevated temperature (> 80 °C) [14].

The previous approach, however, does not work for the preparation of uniform nano-particles with a size of approximately 5 nm, which are required for the toplayer sol coating procedure. The sols in this work were synthesized through two different polymeric sol methods. A well-known method described by e.g. Uhlhorn [15] and de Lange [16,17] was used for making a common SiO$_2$ sol starting from a Si(OC$_2$H$_5$)$_4$ precursor in ethanol. Comparable Zr metalorganic precursors suffer, however, from a much higher reactivity which prevents the controlled synthesis of a similar zirconia sol. The synthesis of sols with a similar particle size is based in this work on the controlled hydrolysis of Zr(n-OC$_3$H$_7$)$_4$ in n-propanol, in the presence of diethanol amine (DEA) as a precursor modifier/polymerization inhibitor. In order to modify the composition of the sol, a titanium (Ti(n-OC$_3$H$_7$)$_3$ or yttrium precursor (Y(n-OC$_4$H$_9$)$_3$) was added as a doping compound before the addition of DEA.

3. Membrane coating

Dip-coating experiments were performed using an automatic dip-coating device, equipped with a holder (vacuum chuck) for square or disk-shaped substrates. In the dip-coating process, sol particles were deposited as a membrane film by contacting the upper-side of the substrate with the coating liquid, while an under-pressure was applied at the back-side.

SiO$_2$ toplayers and toplayers made of (doped) ZrO$_2$ were fired in air at 400°C, 500°C or 600°C for 2 h and a double coating – calcination cycle was used. Intermediate γ-Al$_2$O$_3$ layers were fired at 600°C. For characterization of the material properties (pore size (N$_2$-adsorption/desorption), phase structure (XRD)), unsupported gel-layers were made by drying the remaining coating liquid in Petri-dishes.

4. Gas permeation measurements

Gas permeation testing was carried out using custom-made stainless steel modules, designed for 39 mm disk-shaped membranes. The membranes were sealed using Viton O-rings, which allows measuring at temperatures up to 200°C.The gas flow through the module was controlled by a mass flow controller (Brooks pressure controller 5866) and for measuring the gas flow at the permeate side of the module a gas flow meter was used (Brooks smart mass flow). In a typical series of measurements, the permeation rate of different gases, including He, H$_2$, CO$_2$ and N$_2$ (Linde gas) was determined, starting from the one with the smallest kinetic diameter and using an equilibration time of a few hours. Then, the next gases were applied and after flushing, each gas was permeated untill equilibration occured. All permeation testing was carried out at 200°C and a pressure of 2.5 or 4 bar.

RESULTS

1. Mesoporous intermediate layers

A variety of materials have been considered for making mesoporous membranes, such as Al$_2$O$_3$, ZrO$_2$, TiO$_2$, SiO$_2$ and CeO$_2$. Because of their specific properties, transition Al$_2$O$_3$ membranes (XRD phase γ-Al$_2$O$_3$) are however until now almost exclusively used as an intermediate membrane layer for the development of a gas separation membrane.

In our lab, it was also experienced that this material shows excellent properties for the fabrication of the envisaged gas separation membranes. The unique property of these layers compared with other mesoporous layers under investigation involves the combination of a pore size below 5 nm and an allowable layer thickness of several micrometers. Figure 1 shows micrographs of such a γ-Al$_2$O$_3$ membrane which was prepared by a double dip-coating – calcination step.

While the deposition of a relatively thick crack-free membrane layer is easy for this kind of material, much remains to be done before such membranes become defect-free. In fact, the most important aspect in the preparation method of these layers is to avoid the frequently obtained defects associated with the dip-coating procedure. This is accomplished in our lab by cleaning the surface of the supports, filtering the coating liquid with a 0.8 μm membrane filter and removing air bubbles from the coating liquid. Further, the effect of dust particles is avoided by working in a cleanroom. In the overview surface micrograph in Figure 1b, the effectiveness of this procedure is confirmed by the lack of any defects in the γ-Al$_2$O$_3$ membrane layer.

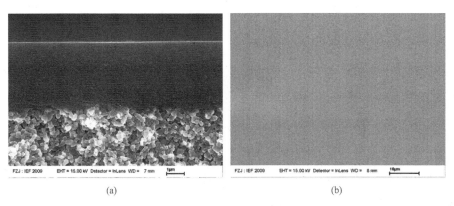

(a)   (b)

Fig. 1. Micrographs of a mesoporous γ-Al$_2$O$_3$ membrane, dip-coated on a macroporous Al$_2$O$_3$ support with a pore size of ~ 100 nm (particle size in sol ~ 30 nm)
((a) cross-section micrograph, (b) overview surface micrograph)

2. Silica and non-silica membrane toplayers

2.1 Morphological properties

The synthesis methods for non-silica toplayers suffer from two major differences when compared with the silica route, namely the high reactivity of the metalorganic precursors in comparison with Si precursors – requiring stabilization with precursor modifiers – and the smaller critical layer thickness. A few examples of membranes with ZrO$_2$, TiO$_2$ and TiO$_2$-ZrO$_2$ toplayers were already shown in reference nr. 18. These toplayers were prepared starting from sols with a molar ratio Zr- and Ti-n-propoxide:DEA in the range 1.5 to 2 and an alkoxide:water ratio in the range 5 to 7. Here, Figures 2a-c show in addition pictures of a membrane with an 8Y$_2$O$_3$-ZrO$_2$ toplayer which is prepared in a similar way. In Figure 2b, an image taken in the back-scattering mode is shown. In Figure 2d, a SiO$_2$ toplayer coated on a similar γ-Al$_2$O$_3$ carrier is also presented.

Comparison of all these different toplayer materials showed several interesting trends. First, the layer thickness decreases in the order amorphous SiO$_2$ >> amorphous TiO$_2$-ZrO$_2$ > crystalline 8Y$_2$O$_3$-ZrO$_2$ and ZrO$_2$. This order also corresponds to the difficulty level of the preparation procedure.

2.2 Gas Permeation tests

Table 1 summarizes the He, H$_2$, CO$_2$ and N$_2$ test results of membranes with silica and zirconia toplayers, dip-coated on mesoporous γ-Al$_2$O$_3$ sublayers. As can be seen, it was found that the gas permeation properties of both toplayers differ dramatically.

Membranes with SiO$_2$ toplayers show a behaviour which is in agreement with the micropore diffusion model, with a higher He permeation than H$_2$ permeation. For N$_2$, no permeation was found, which means that all samples have a selectivty of 100% for H_2 towards N_2. For CO_2, two samples showed no permeation, which means that these membranes have also a selectivity of 100% for H_2 towards CO_2. The three other samples showed a very low CO_2 permeation and a separation factor in the range 10 – 25, while the H_2/CO_2 separation factor of the γ-Al_2O_3 sublayer measures approximately 4.

The observed gas permeation results ($F_{He} > F_{H2} > F_{CO2} > F_{N2}$) and exceptional selectivity of the SiO_2 membrane were somewhat in line with expectations. Similar membranes reported in literature show a similar behaviour and have yielded also relatively high separation factors. By optimising all conditions in the membrane manufacturing procedure in our lab (e.g. support, sol, cleanroom coating) a selectivity of 100% could obtained for a few samples. Currently, we are furter optimizing all parameters in our preparation procedure in order to further improve the reproducibility of our membranes.

By comparing our results with those in the literature, it can also be concluded that the H_2 permeation of the membranes in this work is quite low. The difference can be understood partly by the difference of the toplayer thickness and also the influence of the sublayers should not be neglected, but even then this effect is quite unusual. The permeation of the SiO_2 membrane after firing at 800°C is also listed in Table 1. It can be seen that this 'high temperature' SiO_2 membrane excludes also H_2 and shows only a relatively small He flow.

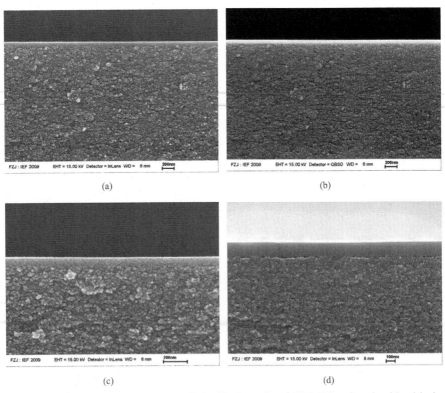

(a)

(b)

(c)

(d)

Fig. 2a. Micrograph of a multilayer-membrane, obtained by dip-coating an $8Y_2O_3$-ZrO_2 polymeric sol (particle size ~ 6 nm) on the mesoporous membrane shown in Fig.1. Fig. 2b. Back-scattering micrograph. Fig. 2c. Detail micrograph. Fig. 2d. Detail micrograph of a SiO_2 membrane, obtained by dip-coating a SiO_2 polymeric sol (particle size ~ 7 nm) on the mesoporous membrane shown in Fig.1.
(Layers made by 2 x dip-coating and calcination at 450°C or 500°C (SiO_2))

Table 1. Permeation data of membranes with SiO_2 and ZrO_2 toplayers

Membrane – Calcination temperature	ΔP (bar)	He ($l/h.m^2.bar$)	H_2 ($l/h.m^2.bar$)	CO_2 ($l/h.m^2.bar$)	N_2 ($l/h.m^2.bar$)	H_2/CO_2
Al_2O_3 Substrat + mesoporous γ-Al_2O_3 interlayer	0.65	8084	11273	2566	3175	4.39
SiO_2-S1 500°C	4	716	382	16.7	0	23
SiO_2-S2 500°C	4	687	168	0	0	100%
SiO_2-S3 500°C	4	630	138	0	0	100%
SiO_2-S4 500°C	2.5	817	216	11.5	0	18.78
SiO_2-S5 500°C	2.5	927	225	18.3	0	12.30
SiO_2-S6 800°C	2.5	68.8	0	0	0	
SiO_2-S6 800°C	4	74.0	0	0	0	
10% Al_2O_3-SiO_2 500°C	4	621	143	0	0	100%
10% TiO_2-SiO_2 500°C	4	1433	286	22.4	0	12.77
3% ZrO_2-SiO_2 500°C	2.5	917	717	53.5	9.9	13.40
20% NiO-SiO_2 500°C	4	1767	1051	37.7	0	27.88
20% Co_3O_4-SiO_2 500°C	4	1194	530	6.7	0	79.10
ZrO_2-S1 400°C	2.5	0	0	0	0	
ZrO_2-S1 500°C	2.5	0	4.7	0	0	
ZrO_2-S1 600°C	2.5	71.1	81.0	14.5	16.8	
ZrO_2-S1 400°C	4	7.8	4.7	0	0	
ZrO_2-S1 500°C	4	12.4	14.3	0	0	
ZrO_2-S1 600°C	4	66.4	83.6	18.6	29.0	
ZrO_2-S2 400°C	2.5	11.4	17.6	0	0	
ZrO_2-S2 400°C	4	22.4	38.2	4.7	0	
ZrO_2-S3 400°C	2.5	9.1	12.4	0	0	
ZrO_2-S3 400°C	4	21.9	28.2	4.7	0	
$8Y_2O_3$-ZrO_2-S1 400°C	2.5	3.3	5.2	0	0	
$8Y_2O_3$-ZrO_2-S1 500°C	2.5	0	3.8	0	0	
$8Y_2O_3$-ZrO_2-S1 600°C	2.5	8.4	8.4	0	0	
$8Y_2O_3$-ZrO_2-S1 400°C	4	11.9	15.3	0	0	
$8Y_2O_3$-ZrO_2-S1 500°C	4	7.2	9.1	0	0	
$8Y_2O_3$-ZrO_2-S1 600°C	4	13.4	8.6	0	0	
$8Y_2O_3$-ZrO_2-S2 400°C	2.5	0	0	0	0	
$8Y_2O_3$-ZrO_2-S2 400°C	4	2.4	4.8	0	0	
$8Y_2O_3$-ZrO_2-S3 500°C	4	23.9	-	-	-	
$8Y_2O_3$-ZrO_2-S4 500°C	4	4.7	-	-	-	
$8Y_2O_3$-ZrO_2-S5 600°C	4	0	-	-	-	
TiO_2-ZrO_2-S1 400°C	2.5	0	0	0	0	
TiO_2-ZrO_2-S1 500°C	2.5	0	0	0	0	
TiO_2-ZrO_2-S1 600°C	2.5	0	0	0	0	
TiO_2-ZrO_2-S1 400°C	4	0	0	0	0	
TiO_2-ZrO_2-S1 500°C	4	0	0	0	0	
TiO_2-ZrO_2-S1 600°C	4	0	0	0	0	
TiO_2-ZrO_2-S2 400°C	2.5	2.3	4.6	0	0	
TiO_2-ZrO_2-S2 500°C	2.5	9.9	15.3	-	-	
TiO_2-ZrO_2-S2 600°C	2.5	16.8	22.9	2.3	-	
TiO_2-ZrO_2-S2 400°C	4	7.2	9.5	0	0	
TiO_2-ZrO_2-S2 500°C	4	13.4	19.1	1.6	3.8	
TiO_2-ZrO_2-S2 600°C	4	25.8	35.3	5.7	-	

S1, S2, S3, ..: Sample 1, Sample 2, Sample 3, ...
- : no data

Contrary to the standard low temperature SiO_2 membranes, measurements on all non-silica toplayers showed no or a negligible He and H_2 permeation, which means that the toplayer also excludes these gases. This result was remarkable since particle size measurements on both silica and zirconia based sols gave values in the same range (5 - 10 nm), the coating method was the same and the thickness of the non-silica toplayers was in all cases smaller.

In Table 1, it can also be seen that the gas permeation results are in most cases in agreement with the Knudsen equation with $F_{He} < F_{H2}$ ($M_{He} = 4$ g/mol, $M_{H2} = 2$ g/mol). The negligible gas flow can therefore be ascribed to the presence of a few inevitable small defects in some samples. It appeared also that the He and H_2 permeation increased by increasing the trans-membrane pressure from 2.5 to 4 bar, which supported the previous conclusion. Further, in Table 1 it can also be seen that the membranes exclude CO_2 and N_2 in almost all measurements.

In order to investigate the possible negative influence of remaining carbon residues from the added precursor modifier (DEA) in the sol-gel coating of non-silica toplayers and to exclude the possible effect of pore blocking, samples with an increasing firing temperature of 400°C, 500°C and 600°C were compared. For most samples however – especially for the $8Y_2O_3$-ZrO_2 toplayers – only some fluctuations were observed, suggesting that the extremely low gas permeation was not arising from a pore blocking effect. For the ZrO_2 toplayer fired at 600°C, a more significant increase of the gas permeability was observed. The latter effect can be understood as arising from a grain growth effect. Figure 3a shows this effect, which can be accompanied by the formation of a few larger pores. XRD on similar toplayers coated on a Si-wafer showed also clearly the formation of the monoclinic phase after firing at 600°C (Figure 3c). However, it can be seen in Table 1 that in this case the gas permeation data were also in agreement with Knudsen diffusion.

For the $8Y_2O_3$-ZrO_2 toplayer (Figure 3b), this grain growth effect was not observed and membranes with such a toplayer showed also after firing at 600°C the same extremely low He and H_2 permeation.

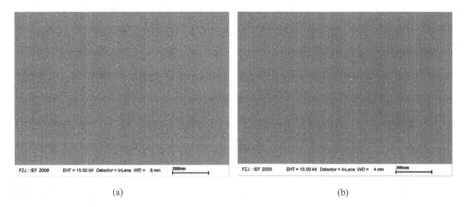

(a) (b)

Fig. 3a. Surface micrograph of a ZrO_2 toplayer, dip-coated on the mesoporous γ-Al_2O_3 layer shown in Figure 1.
Fig. 3b. Surface micrograph of an $8Y_2O_3$-ZrO_2 toplayer, dip-coated on the mesoporous γ-Al_2O_3 layer.
(firing 600°C, particle size in polymeric sol ~ 6 nm)

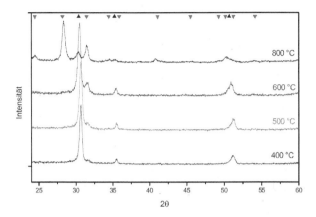

Fig. 3c. XRD of ZrO$_2$ toplayers coated on a Si-wafer

DISCUSSION

The formation of crack-free non-silica membranes was initially experienced as very difficult, but from the provided results, it appears that ultra-thin non-silica membranes with a comparable quality as SiO$_2$ membranes are obtained. The current hurdle which prevents however application of the novel membranes is surprisingly the lack of the required microporosity and therefore a viable gas flow for any gas including H$_2$.

Therefore, after analysis of the previous results, it is suspected that a structural difference of both materials is responsible for the remarkable results. A microporous sol-gel SiO$_2$ membrane shows an amorphous structure. The micro-structure of such a membrane is often described as a 3D network with intramolecular interconnected pores. Most authors assume that more or less linear and weakly branched fractals with a size of several nanometer, which are present in the sol, link together and build a 3D network with residual microporosity. In this point of view, no discrete particles are considered and the sol is often called a 'polymeric sol'. Further, several authors propose that the added water in the sol act as a pore builder or template. In additional XRD experiments, we observed also that the membrane remains its amorphous character also after firing at 800°C.

The main difference between the silica and the zirconia based membranes is that the latter show a clear crystalline structure. For the pure zirconia membrane, the tetragonal polymorph was identified and after the addition of 8 mol% 8Y$_2$O$_3$, the cubic polymorph was found. The TEM pictures presented in Figure 4 confirm the formation of crystalline nano-particles from the sol at a temperature as low as 300°C. In the first micrograph it appears that the 'particles' in the gel state show a mainly amorphous character. From the following micrographs, the formation of crystalline nano-particles with an average size of approximately 5 nm can be observed. The phase structure of these particles corresponds to the cubic polymorph of zirconia, which is in accordance with XRD measurements. Further, it appears that the size of the particles remains similar after further firing in the tested range up to 600°C.

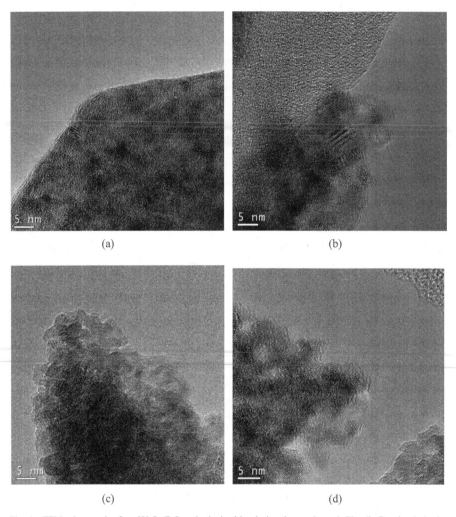

(a) (b)

(c) (d)

Fig. 4a. TEM micrograph of an 8Y$_2$O$_3$-ZrO$_2$ gel, obtained by drying the coating sol. Fig. 4b. Powder derived from the gel fired at 300°C. Fig. 4c. Fired at 400°C. Fig. 4d. Fired at 600°C.

The resulting zirconia toplayers should therefore rather be described as thin films of discrete particles which are linked together. We should expect the occurence of a gas flow through passageways between these particles, but since the required gas permeation is lacking, it seems that these particles form a dense network with no residual porosity even after firing at temperatures as low as 400°C, where sintering effects are not expected.

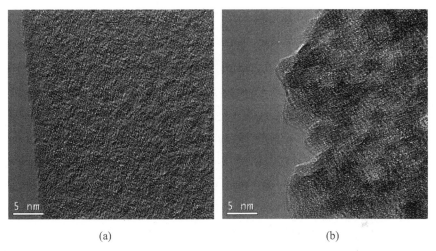

(a)   (b)

Fig. 5a. TEM micrograph of a TiO$_2$-ZrO$_2$ gel (50 mol% - 50 mol%), obtained by drying the coating sol..
Fig. 5b. Powder derived from the gel fired at 500°C.

Based on this result, alternative non-silica toplayers with an amorphous structure were coated and tested for their gas permeation. As can be seen from the TEM micrographs in Figure 5, an amorphous material was obtained by using a mixed Zr:Ti sol with molar ratio 1:1 and this material remains in an amorphous state up to 500°C. In Figure 6a, a toplayer coated on the γ-Al$_2$O$_3$ carrier is also presented. The first tests on this material show however that the smaller gases He and H$_2$ are again excluded to a large extent, which is an indication that the required passageways for these gases are also failing. Then we expand our work to include the high temperature SiO$_2$ membrane fired at 800°C and it appeared that this amorphous membrane also excludes H$_2$ and showed only a limited He permeation.

The obtained results can thus not be ascribed to the amorphous/crystalline character of the silica and non-silica membranes. It appears that some of the tested membrane toplayers posses passageways which allow He and H$_2$ to pass, while other toplayers behave like ultra-thin dense films.

In an attempt to improve the gas permeability of the ZrO$_2$ based membranes, a percentage of Si precursor was also added to the Zr precursor in the synthesis procedure. Figure 6b shows that a similar membrane is obtained from SiO$_2$-doped ZrO$_2$ material (in this example 20 mol% SiO$_2$) but also this type of membrane showed a negligible He permeability of 12 l/h.m^2.bar. On the other hand, by doping an amount of Zr in the SiO$_2$ membrane, membranes with the required permeation were obtained as can be seen in Table 1. Further, similar results were obtained for doping the SiO$_2$ sol with Ti, Al, Ni or Co.

In summary, we can quickly see that among the variety of materials which have been employed for fabricating ultra-thin toplayers, only the SiO$_2$ based toplayers hold currently the potential for application in gas separation. However, this material was in the first place included in this study as a reference material to demonstrate the effectiveness of our sol-gel coating procedures. In order to improve the stability in steam, a variety of additives have been added to silica. The effectiveness of these additions is highlighted in a number of papers and this is also the subject of a part of our work in the future, which will be focused on long-term stability tests with the developed membranes.

However, due to the demonstrated higher stability of non-silica membrane materials such as ZrO$_2$, TiO$_2$ or CeO$_2$ in a number of other membrane applications, it will be equally important to search for (alternative) non-silica sol synthesis and coating methods. In particular, on the long term, it will be important to find a method to prepare toplayers with the required gas flow and which consist of materials such as stabilized zirconia or stabilized ceria which are expected to show the required thermal and chemical stability.

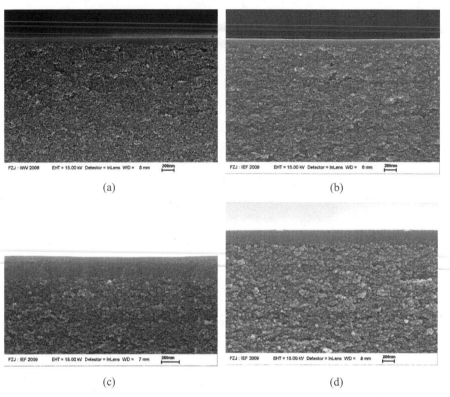

(a) (b)

(c) (d)

Fig. 6a Micrograph of a multilayer-membrane, obtained by dip-coating a TiO$_2$-ZrO$_2$ polymeric sol (50 mol% TiO$_2$).
Fig. 6b Micrograph of a multilayer-membrane, obtained by dip-coating a SiO$_2$-ZrO$_2$ polymeric sol (20 mol% SiO$_2$).
Fig. 6c Micrograph of a multilayer-membrane, obtained by dip-coating a ZrO$_2$-SiO$_2$ polymeric sol (3 mol% ZrO$_2$).
Fig. 6d Micrograph of a multilayer-membrane, obtained by dip-coating a TiO$_2$-SiO$_2$ polymeric sol (10 mol% TiO$_2$)

CONCLUSION

In this paper, silica and non-silica toplayers (ZrO$_2$, 8Y$_2$O$_3$-ZrO$_2$, TiO$_2$-ZrO$_2$) have been prepared on homemade substrates with mesoporous γ-Al$_2$O$_3$ mesoporous interlayers. Based on He and H$_2$ permeability measurements, each of the membranes with non-silica toplayers were classified as gas-tight, while the membranes with a silica toplayer show molecular sieving (e.g. He/N$_2$, H$_2$/N$_2$, H$_2$/CO$_2$).

While in power plants hydro-thermal conditions prevail, it is however necessary to apply membranes which do not degrade in contact with water vapour and there is evidence that SiO$_2$ membranes suffer from serious degradation. At this moment, a few possible application areas are planned for the membrane in the power plant, which include stages with a lower and a higher steam pressure. Improving the steam stability of silica membranes by adding additives may lead to a use of these materials in some stages. A study on the effectiveness of such modifications is currently ongoing in our lab.

Substantial progress has also been achieved in developing novel non-silica membranes for gas separation. In this stage of development, most research efforts were devoted to coating methods of these alternative materials, rather than to the optimization of the (micro)porous structure. Unlike the membranes reported in literature, we have seen here a number of membranes with a very low or no permeability of the larger gases (N$_2$, CO$_2$), which means that our toplayers hold the potential for application in gas separation. Unfortunately, the membranes also lack the required flow for the smaller gases (He, H$_2$). It is not yet known whether we can make non-silica toplayers with the same He or H$_2$ flow as silica layers, but we think that our efforts hold a potential since the sol preparation, the sol particle size and the coating methods are all very similar. To become viable as a separation membrane, the only technical barrier that must be overcome is the creation of a comparable porosity in the toplayers of these membranes.

An extension of the development of non-silica toplayers is the development of a chemically and thermally stable mesoporous carrier. Regardless of which material is used for the toplayer, we suspect that research on these materials will become also an increasingly important factor since the entire multilayer membrane should be able to withstand the operating conditions in the power plant. In our previous work, the potential of stabilized zirconia – which is widely recognized for its excellent chemical and thermal stability – as an alternative for common transition alumina mesoporous interlayers has already been shown.

REFERENCES
[1] Clem E. Powell, Greg G. Qiao, Polymeric CO$_2$/N$_2$ gas separation membranes for the capture of carbon dioxide from power plant flue gases, Journal of Membrane Science 279 (2006) 1–49
[2] R.M. de Vos and H. Verweij, Improved performance of silica membranes for gas separation. J. Membr. Sci. 143 (1998), p. 37
[3] S. Gopalakrishnan, J.C. Diniz da Costa, Hydrogen gas mixture separation by CVD silica membrane, J. Membr. Sci. 323 (2008) 144–147
[4] G. Xomeritakis, C.Y. Tsai, Y.B. Jiang, C.J. Brinker, Tubular ceramic-supported sol–gel silica-based membranes for flue gas carbon dioxide capture and sequestration, J. Membr. Sci. (2009) 30-36
[5] M. Kanezashi, M. Asaeda, Hydrogen permeation characteristics and stability of Ni-doped silica membranes in steam at high temperature, J. Membr. Sci. 271 (2006) 86-93.
[6] S. Battersby, T; Tasaki, S. Smart, B. Ladewig, S. Liu, M.C. Duke, V. Rudolpha, J.C. Diniz da Costa, Performance of cobalt silica membranes in gas mixture separation, J. Membr. Sci. 329 (2009) 91–98
[7] Yoshida, K., Hydrothermal stability and performance of silica-zirconia membranes for hydrogen separation in hydrothermal conditions. J. Chem. Eng. Japan 34(4) (2001) 523-530.
[8] Y. Gu, P. Harcarlioglu, S.T. Oyama, Hydrothermally stable silica-alumina composite membranes for hydrogen separation, J. Membrane Sci. 310 (2008) 28-37
[9] Renate M. de Vos, Wilhelm F. Maier, Henk Verweij, Hydrophobic silica membranes for gas separation, J. Membr. Sci. 158 (1999) 277-288

[10] J.F. Vente, J. Campaniello, C.W.R. Engelen, W.G. Haije, P.P.A.C. Pex, Long Term Pervaporation Performance of Microporous Methylated Silica Membranes, Proceedings of the Eighth International Conference on Inorganic Membranes (ICIM 8) (2004) pp. 81-84

[11] Hessel L. Castricum, Robert Kreiter, Henk M. van Veen, Dave H.A. Blank, Jaap F. Vente, Johan E. ten Elshof, High-performance hybrid pervaporation membranes with superior hydrothermal and acid stability, J. Membr. Sci. 324 (2008) 111-118

[12] M. Hong, S. Li, J.L. Falconer, R.D. Noble, Hydrogen purification using a SAPO-34 membrane, J. Membr. Sci. 307 (2008) 277-283

[13] F. Bonhomme, M.E. Welk, T.M. Nenoff, CO_2 selectivity and lifetimes of high silica ZSM-5 membranes, Micr. Mesop. Mat. 66 (2003) 181–188

[14] B.E. Yoldas, Preparation of glasses and ceramics from metal-organic compounds, J. Mat. Sci. 12 (1977) 1203-1208

[15] R. J. R. Uhlhorn, K. Keizer, A. J. Burggraaf, Gas transport and separation with ceramic membranes. Part II. Synthesis and separation properties of microporous membranes J. Membr. Sci. 66 (1992) 271-287

[16] R.S.A. de Lange, J.H.A. Hekkink, K. Keizer, A.J. Burggraaf, Polymeric-silica-based sols for membrane modification applications: sol-gel synthesis and characterization with SAXS, J. Non-Cryst. Solids 191 (1995) 1-16

[17] R. S. A. de Lange, J. H. A. Hekkink, K. Keizer, A. J. Burggraaf, Formation and characterization of supported microporous ceramic membranes prepared by sol-gel modification techniques, J. Membr. Sci. 99 (1995) 57-75

[18] T. Van Gestel, D. Sebold, H. Kruidhof, H.J.M. Bouwmeester, ZrO_2 and TiO_2 membranes for nanofiltration and pervaporation: Part 2. Development of ZrO_2 and TiO_2 toplayers for pervaporation, J. Membr. Sci. 318 (2008) 413-421

THERMAL SHOCK PROPERTIES OF POROUS ALUMINA FOR SUPPORT CARRIER OF HYDROGEN MEMBRANE MATERIALS

Sawao HONDA[1], Yuuki OGIHARA[1], Shinobu HASHIMOTO[2] and Yuji, IWAMOTO[1]

[1]Department of Frontier Materials, Graduate School of Engineering, Nagoya Institute of Technology
[2]Department of Materials Science and Engineering, Graduate School of Engineering, Nagoya Institute of Technology

466-8555 Gokiso-cho, Showa-ku, Nagoya, Aichi 466-8555, Japan

ABSTRACT

The thermal shock resistance of α-alumina porous capillary, the support material for hydrogen-permselective microporous ceramic membrane was studied. To study the effect of porosity on the thermal shock resistance systematically, porous alumina with different porosities was fabricated, and the thermal shock resistance of the fabricated samples as well as the porous capillary was estimated by the infrared radiation heating method. The mechanical and thermal properties concerned to the thermal shock resistance were also measured and the effect of the porosity on the properties was carefully examined. The fracture strength was not changed with temperature, but decreased with the porosity. The fracture toughness, Young's modulus and thermal conductivity were also decreased with porosity. Thermal shock resistance of porous alumina was estimated quantitively by the experimental thermal shock parameters, thermal shock strength, R_{1c} and thermal shock fracture toughness, R_{2c}. The thermal shock parameters of porous alumina were much lower than dense alumina, and decreased with porosity due to the decreasing of fracture strength and thermal conductivity. The experimental thermal shock strength was good accordance with that calculated from the material properties in this study. Thermal shock strength of porous alumina capillary at service temperature could be estimated by the comparison experimental and calculated thermal shock strength.

INTRODUCTION

Porous ceramics has been used for environmental and energy industries. In hydrogen (H_2) production process, for example methane (CH_4) steam reforming reaction, H_2 gas can be produced at approximately 1073 K, and must be separated from other mixed gas using palladium-based membranes or microporous ceramic membranes. Microporous ceramic membranes have relatively high gas permeability and good stability at high temperature. Porous alumina capillary has been investigated as support material for the ceramic membranes.[1] Generally, the porous capillary support has fine pores (about 0.1μm) and relatively large porosity (above 40 %). This porous structure design is essential for the gas separation membrane support to minimize the pressure drop of the permeating gas. However, thermal shock fracture will be occurred by the thermal stress which is caused by a rapid temperature change realized to the reaction temperature, or unexpected emergency stop of the H_2 production reactor. The estimation of thermal shock resistance is important to the application of porous alumina capillary to ceramic membrane for the H_2 production process.

In past studies, many researches of the mechanical and thermal properties of porous ceramics have been reported. The limited literatures, however, exist on the estimation of the thermal shock resistance properties for the porous ceramics, especially high porosity ceramics.[2)-4)] The quantitative estimation of the relation thermal shock resistance and porous structure is insufficient. The water quenching mainly has been used for the estimation of thermal shock resistance. In the case of application to the porous ceramics, however, the penetration of quenching medium through the pores and instability of heat transfer coefficient at the rough surface were afraid to cause difficulty in the accurate estimation. On the other hand, we have been evaluated thermal shock resistance of the dense ceramics by the infrared radiation heating (IRH) method.[5]-[8] It has advantage that can be stable heat flux condition and quantitative estimation of thermal shock resistance using thermal shock parameters.

In this study, we investigate to apply the IRH method to the high porosity ceramics as thermal shock testing method. Porous alumina with fine pore size and different porosities are fabricated by change of the sintering temperature. Thermal shock resistance is estimated by the IRH method, and the mechanical and thermal properties concerned to the thermal shock resistance are also measured. The relations between these properties and porosity are evaluated by thermal shock parameters, and the thermal shock resistance of porous alumina capillary is estimated quantitively.

EXPERIMENTAL PROCEDURE

Porous alumina preparation

The specimen sizes used in this study are listed in Table 1. Preparation of porous alumina capillary specimen (CS) was reported by KOJIMA, et. al.[1] Because of the shape and size of the capillary was not suitable for the specimen of thermal shock resistance and some other properties, bulk specimen was fabricated by the powder forming process in this study (PS). The green tubes consisted of α-alumina powder (TAIMEI Chemicals Co. Ltd., TM-DA, 0.3 μm in mean particle size) and polymer binder was prepared by a dry-wet spinning method, which was grinded using the mortar. After this grinded powder was passed through a 60 mesh sieve, it was pressed by uniaxially load (0.4 MPa) using disk shaped dies with the diameter 20 or 50 mm. These green disks were sintered by partial sintering method in the temperature at 1423 (PS14), 1523 (PS15) and 1623 (PS16) K for 30 min in air. By changing the sintering temperature, the porosity of the porous alumina was controlled. Another porous alumina specimen with different pore size was prepared by sheet forming process (SS) using same grinded powder, sintered at 1423 K. All of the sintered bodies were cutting with diamond saw to the various specimen sizes.

Mechanical and thermal properties

The density and porosity of the porous alumina were measured by the Archimedes method. The mean pore diameter was determined by a mercury porosimeter (Shimadzu Co., PORE SIZER 9320). The total porosity of the porous alumina was calculated as:

$$P_t = P_{op} + \left(1 - P_{op}\right)\left(1 - \frac{\rho_a}{\rho}\right) \tag{1}$$

where P_t is total porosity, P_{op} open pore porosity, ρ_a apparent density, and ρ theoretical density. The fracture strength, fracture toughness, and Young's modulus were evaluated by a universal testing machine with electric furnace (Instron Co., 5582) at a cross head speed of 1.0 mm/min. The fracture strength of CS, σ_C, was calculated as:

Table 1 Experimental conditions and specimen sizes in the measured properties.

Property	Sample name	Method	Temperature (K)	Specimen shape	Specimen size (mm)
Fracture strength	CS	Three-point bending (Lower span 30 mm)	R.T., 873, 1023	Cappilry	2.8x2.5x40 (outer x inner x length)
	PS	Three-point bending (Lower span 8 mm)	R.T.	Rectangular Bar	2x2x10
Fracture toughness	PS	SEVNB (Lower span 8 mm)	R.T.	Rectangular Bar	2x2x10 (Notch length 1.0)
Young's modulus	PS	Commpression test with strain-gage	R.T.	Rectangular Bar	5x5x10
Thermal conductivity	PS	Laser flash	R.T., 473, 673 873, 1023	Disk	□10×1
Thermal expansion coefficient	PS	Differntial thermal expansion meter	R.T., 473, 673 873, 1023	Rectangular Bar	5x5x10
Thermal shock resistance	PS, SS	Infrared radiation heating	R.T. (Initial temp.)	Disk	□37-40 $(2r_0)$ ×1,1 (H) $(c = 2.0)$

$$\sigma_C = \frac{8 D_2 L P_f}{\pi \left(D_2{}^4 - D_1{}^4 \right)}$$

(2)

where P_f is fracture load, D_2 outer diameter, D_1 inner diameter, and L lower span. On the other hand, the fracture strength test of PS was based on JIS R 1601. Fracture toughness was estimated by the Single edge V-notched beam (SEVNB) method.[9] The specimens similar to those for bending test in dimensions were introduced at the lower surface with a sharp V-notch. The radius of curvature at the V-notch tip was about 20 μm. The fracture toughness, K_{IC}, test was based on JIS R 1607, and it was calculated as:

$$K_{IC} = \frac{3}{2} \frac{P_f L}{B V^{3/2}} Y \sqrt{\xi} \qquad (\xi = \frac{\gamma}{V})$$

(3)

$$Y = \frac{1.99 - \xi(1-\xi)(2.15 - 3.93\xi + 2.7\xi^2)}{(1+2\xi)(1-\xi)^{3/2}} \qquad \text{(at } L/T = 4.0\text{)}$$

where B is thickness of specimen, V width of specimen, Y shape factor, and γ notch length. Young's modulus was estimated by the slope of the stress-strain curve on the compression testing. Strain was measured by the strain gage attached to a normal surface to the compression surface. Thermal conductivity was measured by the laser flash thermal constants analyzer (Ulvac Riko Co., TC-7000) based on JIS R 1611, and the coefficient of thermal expansion was determined with a differential thermal expansion meter (Shimadzu Co., TMA-50) based on JIS R 1618. The temperature dependences of these thermal properties were also studied by changing the measured temperatures, from R.T. to 1023 K. All of the values in properties of dense alumina $(P=0)$, σ_0, E_0, K_0 and λ_s were used for the values of previous work.[7]

Thermal shock test by Infrared radiation heating

     Disk-shaped specimens were prepared for the thermal shock strength, R_{1c} test, and a V-notch (notch length $c=2.0$ mm) was produced on the disk edge for the thermal shock fracture toughness, R_{2c} test, as showed in **Fig. 1**. Two thermocouples were attached symmetrically to the disk surface at 10 mm from the center of the disk, to provide for estimation of the heating efficiency and the temperature at the fracture point. Graphite was coated onto the heating areas ($2a=14.5$ mm) on both sides of the disk to enhance absorptivity of the infrared rays. The apparatus used for the IRH method was represented in **Fig. 2**. The surface of the specimen disk was covered with heat insulating material, to prevent infrared radiation except in the heating area and heat loss by air convection around the specimen. The infrared rays were radiated from the two lamps at both sides of the specimen, through the ellipsoid mirror and quartz bars. After the supplied electric powers to the lamps were obtained to setting values, the center of the disk was irradiated by opening an aluminum shutter. The thermal shock fracture was occurred by

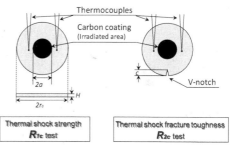

Fig.1 Schematic diagram of disk specimens in infrared radiation heating method.

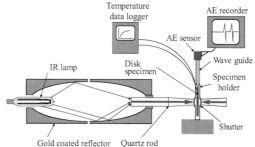

Fig.2 Schematic diagram of thermal shock test by infrared radiation heating method.

the thermal stress generated from temperature difference between the center and circumference of the disk. Start time of crack propagations was determined with an acoustic emission (AE) device through the wave guide. The maximum tensile thermal stress and stress intensity factor at the V-notch tip in specimen were calculated numerically at the failure time, t, determined by AE signals.

Thermal Shock Parameters

Several thermal shock resistance parameters were suggested in the past studies.[10-11] For standardized thermal shock testing, however, thermal shock resistance parameters as material constants should be established on the basis of the physical properties of materials. We proposed two thermal shock parameters as physical properties: thermal shock strength, representing the resistance to thermal shock fracture, and thermal shock fracture toughness, denoting the resistance to the initiation of crack propagation.[12] These two parameters are defined as

$$R_{1c} = \frac{\lambda \sigma_f}{E\alpha} \qquad [\mathrm{W \cdot m^{-1}}] \tag{4}$$

$$R_{2c} = \frac{\lambda K_{IC}}{E\alpha} \qquad [\mathrm{W \cdot m^{-0.5}}] \tag{5}$$

where σ_f is the modified strength estimated from fracture strength of the material with volume effect calculation, K_{IC} is the fracture toughness of the material, E is the Young's modulus, α is the thermal expansion coefficient, λ is the thermal conductivity. Equations (4) and (5), which are defined combinations of the material properties, are referred to herein as 'calculated values' of the thermal shock parameters.

The thermal shock parameters can be evaluated directly from the electrical charge and the testing conditions of the IR heating technique by equations (6) and (7). [12]

$$R_{1c} = S\frac{\eta\omega}{\pi H(a/r_0)^2} \qquad [\mathrm{W \cdot m^{-1}}] \tag{6}$$

$$R_{2c} = N_I\sqrt{\pi c}\frac{\eta\omega}{\pi H(a/r_0)^2} \qquad [\mathrm{W \cdot m^{-0.5}}] \tag{7}$$

where η is the efficiency of the transforming electric power into heat flux, ω the supplied electronic power to the IR ramps, and a the radius of the heated area on the disk, H the thickness of the disk, r_0 the radius of the disk, S the nondimensional thermal stress, and N_I the nondimensional stress intensity factor at the notch tip. In the specimen of the R_{2c} test, K_I and c correspond to the stress intensity factor at the V-notch tip and c the V-notch length, respectively. Equations (6) and (7), defined combinations of the experimental conditions, are referred to herein as 'experimental values' of the thermal shock parameters. The changes with time of S and N_I are calculated from the transient temperature distribution of the disk specimen using the data of thermocouple. The critical values of S and N_I are those of at the fracture time, t. The calculated and experimental values of thermal shock parameters on the dense ceramics were in good agreement for the thermal shock test using the IRH.[6-7]

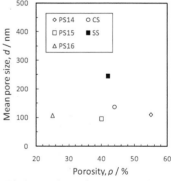

Fig.3 Porosity and mean pore sizes of tested porous alumina.

RESULTS AND DISCUSSION

Properties of porous alumina

Figure 3 shows total porosity and mean pore size (d) of sintered porous alumina. The porosity of PS decreased with an increasing sintering temperature. The mean pore sizes, however, were almost same among PS series. The sintered porous alumina capillaries, which were fabricated by the dry-wet spinning method, also showed the different porosity with same pore diameter of approximately 0.1 μm by changing sintering temperature.[1] The effect of the porosity on the properties of porous alumina could be evaluated by the comparison of PS series. On the other hand, as the porosity of CS, SS, and PS15 were almost same, the effect of pore size could be estimated. **Figure 4** shows a typical SEM image of the as-sintered surface in PS14. The alumina grain and their necking did not grow at this sintering temperature.

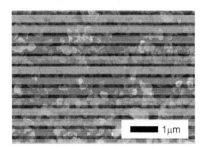

Fig.4 SEM image of the as-sintered surface in porous alumina of PS14.

Figure 5 shows the temperature dependence of the three-point bending strength, σ_C, in CS. The mean strength of σ_C was slightly decreased with an increasing temperature. The fracture strength at 1073 K was decreased about 7 % compared to that of room temperature. The Weibull analysis for the fracture strength at room temperature was carried out. The Weibull modulus (m=15.2) of the CS agreed with the results of porous alumina tube in other research.[1,13] A scatter of fracture strength seemed to be increased with increasing temperature. Nishikawa et al.[14] reported that the fracture strength of 1073 K was decreased about 10 % compared to that of room temperature in porous alumina (porosity of 40%, pore size of 1.5-10 μm). This result showed that the effect of the mean pore size on the temperature dependency of fracture strength was not so large.

Figure 6 shows the relation of the porosity and three-point bending strength in PS, σ_P, and σ_C at room temperature. The values of σ_P were decreased with an increasing porosity. The σ_C was agreed with the relation between σ_P and porosity. Many researches have reported to represent the relationship between porosity (P) and fracture strength (σ). The following general equation was proposed by Ryshkewitch[15] and Knudsen:[16]

$$\sigma = \sigma_0 \, exp\left(-bP\right) \qquad (8)$$

where σ_0 is fracture strength at P=0, and b an empirical constant estimated from slope of semi log plot. The b value of PS (5.2) was agreed with that of the other research in porous alumina (5.1).[17] Eudier suggested the following equation by a model of minimum cross section area of matrix.[18]

$$\sigma = \sigma_0 \left(1 - KP^{2/3}\right) \qquad (9)$$

where $K = \pi \, (3/4\pi)^{2/3}$. As compared the two equations, it was seemed to fit the Eq.(9). The other

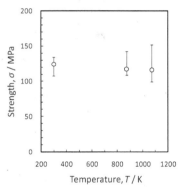

Fig.5 Temperature dependence of fracture strength in CS.

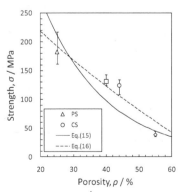

Fig.6 Porosity dependence of fracture strength in PS, and comparison with CS.

research of fracture strength in porous alumina with sub-micron pore size fabricated by the partial sintering method was agreed with the Eq. (9).[19]

Figure 7 shows the relation of the porosity and Young's modulus, E, in PS at room temperature. The E, were decreased with an increasing porosity as well as fracture strength. The E value of PS agreed with porous alumina having sub-micron[19-21] or micron pore size[19],[22] in the same porosity. Young's modulus was slightly depended on the pore size.[19],[22] Many equations had been reported to represent the relationship between porosity (P) and Young's modulus (E). The following equation was proposed by Spprigs[23] and Knudsen:[24]

$$E = E_0 \, exp\left(-bP\right) \quad (10)$$

where E_0 is Young's modulus at $P=0$, and b an empirical constant. The b value of PS calculated from eq. was 6.3. As compared with other porous alumina with the sub-micron $(6.5)^{[19]}$ and micron $(4.0\text{-}4.3^{[19]}, 4.0^{[24]})$ pore size, the b values of porous alumina with sub-micron pore size were higher than that of micron pore size. This results were attributed to lower E of porous alumina with sub-micron pore size in high porosity regions. Because the sub-micron porous alumina was sintered at lower sintering temperature, lower Young's modulus was appeared by insufficient neck growth of alumina grains[19], as showed in Fig. 4. The following equation was suggested for the Young's modulus of partial sintering porous alumina.[20-21]

$$E = E_0 \left(1 - \frac{P}{P_0}\right)^n \quad (11)$$

where E_0 is Young's modulus at $P=0$, P_0 a porosity of the green bodies of the specimen (about 65 % in PS), and n an empirical constant. The n value of PS (1.38) was similar to the other study $(1.35)^{[20]}$.

Figure 8 shows the relation of the porosity and fracture toughness, K_{IC}, in PS at room temperature. The K_{IC}, were decreased with an increasing porosity. The following equation was suggested for the fracture toughness of partial sintering porous alumina as well as Young's modulus.[20-21]

$$K = K_0 \left(1 - \frac{P}{P_0}\right)^n \quad (12)$$

where K_0 is fracture toughness at $P=0$, and n an empirical constant. The n value of PS (1.13) was similar to the other study $(1.15)^{[20]}$.

Figure 9 shows the temperature dependences of thermal conductivity, λ, in PS. The λ were decreased with an increasing temperature as well as dense alumina. They were also decreased with an increasing

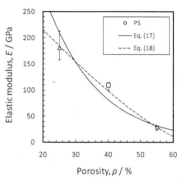

Fig.7 Porosity dependence of Young's modulus in PS.

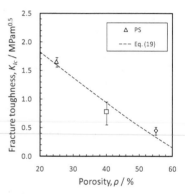

Fig.8 Porosity dependence of fracture toughness in PS.

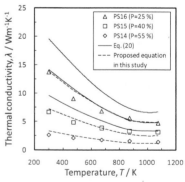

Fig.9 Temperature dependence of thermal conductivity in PS.

porosity. Many equations have been reported to represent the relationship between porosity (P) and thermal conductivity (λ). The following equation was proposed by Russel:[25]

$$\lambda = \lambda_s \frac{P^{2/3} + \kappa\left(1 - P^{2/3}\right)}{P^{2/3} - P + \kappa\left(1 - P^{2/3} + P\right)} \qquad \kappa = \frac{\lambda_s}{\lambda_f} \qquad (13)$$

where λ_s is thermal conductivity of solid part (dense alumina matrix), λ_f thermal conductivity of fluid part (air). Though the calculated values with solid lines were compared with experimental plot in Fig. 9, these values were not agreed with the experimental values. The λ in this study was apparently lower than the values estimated by these equations and the other research.[25)-26] Because of the densification of alumina matrix was not progressed with lower sintering temperature, as showed in Fig. 4, thermal conductivity of solid part of porous alumina in this study was lower than that of the dense alumina. Therefore, we thought that λ_s was also depended on the porosity, and tried to calculate λ by substituting λ_s the dense alumina by the value calculated by Eq. (13). These modified values with broken lines were agreed approximately with the experimental values.

Figure 10 shows the temperature dependences of thermal expansion coefficient, α, in PS. The α were decreased with an increasing temperature as well as dense alumina. They were not changed with porosity different from the other properties. In literature, alumina refractories with porosity 0.3 and 0.7 showed the same thermal expansion coefficient.[27] This result suggests that the thermal expansion of porous alumina is not affected by the porous characteristics such as porosity. Temperature dependence of α could be well fitted by a cubic equation.

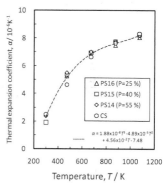

Fig.10 Temperature dependence of in thermal expansion coefficient in PS, and comparison with CS.

Thermal shock test of porous alumina

A photograph of the porous alumina (PS) specimen after the thermal shock strength test was showed in **Fig. 11**. A fracture was occurred at 2-3 mm inside the edge of the specimen disk (indicated by arrow). After the crack propagation to a straight line toward the center of the disk, it was branching and curved. This pattern of the crack propagation was also observed in the thermal shock test using IRH technique for dense ceramics[6]. In the specimen of the higher porosity above 40 %, the crack was propagated in a straight line toward the edge in the opposite side of the disk without branching.

The AE signal detected during the thermal shock test was shown in **Fig. 12**. The start time

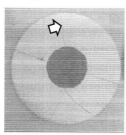

Fig.11 Photograph of fractured specimen after thermal shock strength test.

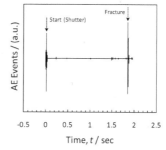

Fig.12 AE signal of thermal shock test.

of heating was decided by the sound of a shutter. In this test, AE signal of the thermal shock fracture was clearly detected after 1.85 seconds from the start of heating. Therefore, the fracture time, t was determined to 1.85 in this test. The numbers of AE events were correlated with the number of cracks or the retained strength of specimen after the thermal shock test.[28-29] In the IRH method, a number of AE events were not changed in each test because only a few large cracks were propagated, as showed in **Fig. 11**. In a porous material, the AE was easily attenuated by a scattering in pores. The thermal shock testing method by water quenching, the AE of the thermal shock fracture was difficult to detect by the noise of boiling water. The IRH method was possible to detect AE clearly without the effect of the noise even when the porous ceramics was tested. Knowledge of the fracture time accurately was enabled to analyze of the thermal stress and the quantitative evaluation of thermal shock resistance. The IRH method was proved to be useful for the thermal shock resistance of the porous ceramics.

Figure 13 shows the relation between experimental R_{1c} of PS obtained from Eq. (6), by thermal stress calculation and various experimental conditions, and calculated R_{1c} derived from Eq. (4), based on measured properties of PS. The R_{1c} decreased with an increasing the porosity. For all the specimens with different porosity, the experimental R_{1c} showed good agreement with the calculated R_{1c}. Thus, the thermal shock strength of porous alumina ceramics could be estimated by their material properties.

Figure 14 shows the relation between experimental R_{2c} of PS obtained from Eq. (7), by calculation of stress intensity factor at the notch and various experimental conditions, and the calculated R_{2c} derived from Eq. (5), based on the measured properties of PS. The experimental R_{2c} showed the larger scatters than R_{1c} in every porosity, and decreased with an increasing the porosity as well as R_{1c}. The experimental R_{2c} was showed not so good agreement with the calculated R_{2c}. In past studies, we reported that the experimental R_{2c} was agreed with calculated R_{2c} on various dense ceramics.[6-7] Thus, this disagreement was caused by the porous specimen. The experimental R_{2c} was higher than the calculated R_{2c} in PS15 and PS16. Because of the disagreement was not appeared in R_{1c}, it was thought to be ascribable to over-estimate the critical value of stress intensity factor by the relaxation of the stress intensity at the notch-tip in thermal shock specimen by compare with the K_{IC} measured by the SEVNB method. If the relaxation was caused by existence of pore, lower porosity specimen with higher inhomogeneity of pore distribution showed the higher scatter of stress intensity factor. Furthermore, the specimen thickness of the thermal shock test was thinner than SEVNB specimen. Notch shape was easily deformed in the thin disk, and this deformation enhanced the relaxation of stress intensity. These effects may be avoided by the adjustment thickness of disk specimen.

A porosity dependence of thermal shock parameters was investigated by the calculated values from Eqs (4), (5). The E, σ, K_{IC}, and λ were decreased, and α was not changed with porosity.

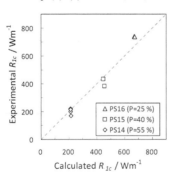

Fig.13 Comparison of experimental and calculated thermal shock strength in PS.

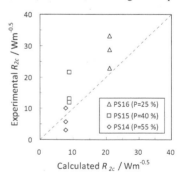

Fig.14 Comparison of experimental and calculated thermal shock fracture toughness in PS.

While the decreasing of E enhanced the thermal shock parameters, the decreasing of σ, K_{IC}, and λ lowered it.

Figure 15 shows the relation between the porosity and the experimental R_{Ic} in PS and SS. The SS with lager pore size than PS showed lower R_{Ic} than PS15 with similar porosity. The fracture strength of porous alumina was depended on the pore size as following equation:[22),30)]

$$\sigma_f \propto \frac{1}{\sqrt{d}} \tag{14}$$

Thus, the fracture strength is decreased with an increasing pore size. The thermal shock strength was decreased with increasing pore size. In order to estimate the thermal shock strength of CS, thermal shock strength was expressed as a function of porosity by substituted for Eqs. (9), (11), (13), and fitted equation of α into Eq.(4). This function showed a relatively good agreement with the experimental R_{Ic} of PS, and the R_{Ic} of CS could be predicted 377 W/m at $P=0.44$. If the material properties are not changed with the fracture strength, the thermal shock strength was also propositional to $d^{0.5}$ from Eq.(14). In this case, the R_{Ic} of CS could be predicted 336 W/m. The thermal shock parameters of dense alumina were evaluated by IRH method in the previous work.[6-7)] These values were estimated 3000 W/m of R_{Ic} and 50 W/m$^{0.5}$ of R_{2c}. The thermal shock resistance of the porous alumina was much lower than that of the dense alumina.

Figure 16 shows the temperature dependences of calculated R_{Ic} in PS. The R_{Ic} was decreased with an increasing temperature due to the decreasing of λ and increasing of α with temperature. The hydrogen-permselective membrane was used at 873-1073 K of reaction temperature. The thermal shock strength of is decreased about 70 W/m at 1073 K. It is important that the porous alumina capillary of thermal shock resistance could be estimated at service temperature.

CONCLUSIONS

The thermal shock resistance of α-alumina porous capillary was estimated by the infrared radiation heating method. The mechanical and thermal properties concerned to the thermal shock resistance were also measured. The fracture strength was not changed with temperature, but it was decreased with the porosity. The fracture toughness, Young's modulus and thermal conductivity were also decreased with porosity. Thermal expansion coefficient was not changed with porosity. Thermal shock test of porous alumina was executed successfully by the thermal shock fracture and AE signals, and the thermal shock resistance of porous alumina could be estimated quantitively by the experimental thermal shock parameters. The thermal shock parameters of porous alumina were much lower than dense alumina, and decreased with porosity due to the decreasing of fracture strength, fracture toughness and thermal conductivity. The experimental thermal shock strength was

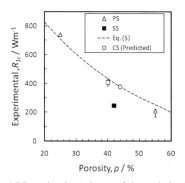

Fig.15 Porosity dependence of thermal shock strength in PS, and comparison with SS and predicted values of CS.

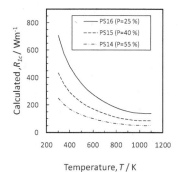

Fig.16 Temperature dependences of thermal shock strength in PS.

good accordance with the calculated thermal shock strength evaluated by the material properties. Thermal shock strength of porous alumina capillary at service temperature could be estimated by the comparison experimental and calculated thermal shock strength.

ACKNOWLEDGEMENT
The authors special thank NOK Corporation for providing the raw material powder and capillary samples. This research was carried out as a part of the R & D project on New Energy and Industrial Technology Development Organization (NEDO), Japan.

REFERENCES
[1] R. Kojima, K. Sato, T. Nagano, and Y. Iwamoto, Development of fine porous alumina capillaries by a dry-wet spinning method, *J. Ceram. Soc. Japan*, **114**, 929-933 (2006).
[2] R. M. Orenstein and D. J. Green, Thermal Shock Behavior of Open-Cell Ceramic Foams, *J. Am. Ceram. Soc.*, **75**, 1899-1905 (1992).
[3] V. R. Vedula, D. J. Green, and J. R. Hellman, Thermal shock resistance of ceramic foams, *J. Am. Ceram. Soc.*, **82**, 649-656 (1999).
[4] J. She, T. Ohji, and Z. Y. Deng, Thermal shock behavior of porous silicon carbide ceramics, *J. Am. Ceram. Soc.*, **85**, 2125-2127 (2002).
[5] S. Honda, T. Takahashi, S. Morooka, S. Zhang, T. Nishikawa, and H. Awaji, Thermal stress and stress intensity factor considering temperature dependent material properties - a circular disk under constant heat flux, *J. Soc. Mat. Sci., Japan*, **46**, 1300-1305 (1997). [in Japanese]
[6] S. Honda, T. Suzuki, T. Nishikawa, H. Awaji, Y. Akimune, and N. Hirosaki, Estimation of thermal shock properties for silicon nitride having high thermal conductivity, *J. Ceram. Soc. Japan*, **110**, 38-43 (2002). [in Japanese]
[7] S. Honda, T. Nishikawa, H. Awaji, N. Hirosaki, and Y. Akimune, Estimation of temperature dependence of thermal shock resistance by infrared radiation technique, *Ceram. Trans.*, **113**, 127-132 (2002).
[8] S. Honda, K. Kimata, S. Hashimoto, Y. Iwamoto, M. Yokoyama, J. Shimano, K. Ukai, and Y. Mizutani, Strength and thermal shock properties of scandia-doped zirconia for thin electrolyte sheet of solid oxide fuel cell, *Mater. Trans.*, **50**, 1742-1746 (2009).
[9] H. Awaji and Y. Sakaida, V-notch technique for single-edge notched beam and chevron notch methods, *J. Am. Ceram. Soc.*, **73**, 3522-23 (1990).
[10] W. D. Kingery, Factors affecting thermal stress resistance of ceramic materials, *J. Am. Ceram. Soc.*, **38**, 3-15 (1955).
[11] D. P. H. Hasselman, Unified theory of thermal shock fracture initiation and crack propagation in brittle ceramics, *J. Am. Ceram. Soc.*, **52**, 600-604 (1969).
[12] H. Awaji, S. Honda, and T. Nishikawa, Thermal shock parameters of ceramics evaluated by infrared radiation heating, *JSME Int.*, **40**, 414-422 (1997).
[13] S. C. Nanjangud, R. Brezny, and D. J. Green, Strength and Young's modulus behavior of a partially sintered porous alumina, *J. Am. Ceram. Soc.*, **78**, 266-68 (1995).
[14] T. Nishikawa, Y. Umehara, S. Honda, and H. Awaji, Mechanical properties of porous alumina at high temperature, *J. Ceram. Soc. Japan, Sup.*, **112**, S1405-1407 (2004).
[15] E. Ryshkewitch, Compression strength of porous sintered alumina and zirconia: 9th communication to ceramography, *J. Am. Ceram. Soc.*, **36**, 65-8 (1953).
[16] F. P. Knudsen, Dependence of mechanical strength of brittle polycrystalline specimens on porosity and grain size, *J. Am. Ceram. Soc.*, **42**, 376-387 (1959).
[17] L. J. Trostel Jr., Strength and structure of refractories as a function of pore content, *J. Am. Ceram. Soc.*, **45**, 563-564 (1962).
[18] M. Eudier, The mechanical properties of sintered low-alloy steel, *Powder Metall.*, **5**, 278-287 (1962).

[19] T. Nishikawa, A. Nakashima, S. Honda, and H. Awaji, Effects of porosity and pore morphology on mechanical properties of porous alumina, *J. Soc. Mat. Sci., Japan*, **50**, 625-629 (2001). [in Japanese]

[20] T. Ostrowski and J. Rodel, Evolution of mechanical properties of porous alumina during free sintering and hot pressing, *J. Am. Ceram. Soc.*, **82** 3080-3086 (1999).

[21] B. D. Flinn, R. K. Bordia, A. Zimmermann, and J. Rodel, Evolution of defect size and strength of porous alumina during sintering, *J. Eur. Ceram. Soc.*, **20**, 2561-2568 (2000).

[22] M. Ashizuka, E. Ishida, T. Matsushita, and M. Hisanaga, Elastic modulus, strength and fracture toughness of alumina ceramics containing pores, *J. Ceram. Soc. Japan*, **110**, 554-559 (2002). [in Japanese]

[23] R. M. Spriggs, Expression for effect of porosity on elastic modulus of polycrystalline refractory material, particularly aluminum oxide, *J. Am. Ceram. Soc.*, **44**, 628-9 (1961).

[24] F. P. Knudsen, Effect of porosity on Young's modulus of alumina, *J. Am. Ceram. Soc.*, **45**, 94-95 (1962).

[25] H. W. Russel, Principles of heat flow in porous insulators, *J. Am. Ceram. Soc.*, **18**, 1-5 (1935).

[26] Z. Zivcova, E. Gregorova, W. Pabst, D. S. Smith, A. Michot, and C. Poulier, Thermal conductivity of porous alumina ceramics prepared using starch as a pore-forming agent, *J. Eur. Ceram. Soc.*, **29**, 347-353 (2009).

[27] J. B. Austin, Thermal expansion of nonmetallic crystals, *J. Am. Ceram. Soc.*, **35**, 243-253 (1952).

[28] K. J. Konsztowicz, Crack growth and acoustic emission in ceramics during thermal shock, *J. Am. Ceram. Soc.*, **73**, 502-508 (1990).

[29] F. Mignard, C. Olagnon, and G. Fantozzi, Acoustic emission monitoring of damage evaluation in ceramics submitted to thermal shock, *J. Eur. Ceram. Soc.*, **15**, 651-653 (1995).

[30] T. Isobe, Y. Kameshima, A. Nakajima, K. Okada, and Y. Hotta, Gas permeability and mechanical properties of porous alumina ceramics with unidirectionally aligned pores, *J. Eur. Ceram. Soc.*, **27**, 53-59 (2007).

IN SITU PROCESSING OF POROUS $MgTi_2O_5$ CERAMICS WITH PSEUDOBROOKITE-TYPE STRUCTURE TOWARD THIRD GENERATION DIESEL PARTICULATE FILTER MATERIALS

Yoshikazu Suzuki

Institute of Advanced Energy, Kyoto University, Gokasho, Uji, Kyoto 611-0011, Japan

ABSTRACT

Double-oxide-based porous ceramics are promising for third-generation DPF materials, following to cordierite (first-generation) and silicon carbide (second-generation). Up to now, mullite ($3Al_2O_3 \cdot 2SiO_2$) and aluminum titanate (Al_2TiO_5) are of particular interest as candidates of the third generation, due to their low coefficients of thermal expansion (CTE) among oxides. Aluminum titanate has a pseudobrookite-type crystal structure, and this structure generally shows the anisotropic thermal expansion due to the distortion of atomic coordination polyhedra. As for other pseudobrookite-type ceramics, however, they are not so widely studied as aluminum titanate (especially in porous forms). In this study, porous $MgTi_2O_5$ ceramics with pseudobrookite-type structure have been successfully prepared by *in situ* processing (*viz.* reactive sintering). Preliminary microstructural observation of the porous $MgTi_2O_5$ will be presented.

INTRODUCTION

Diesel particulate filters (DPFs) are widely used for trapping particulate matter (PM) in the diesel exhaust.[1-4] High-performance DPFs have been mass-produced by several ceramic-related companies, *e.g.*, NGK, Ibiden and Corning. Figure 1 shows sales trends for DPF-related products of major DPF suppliers. Although the production has declined in 2008-2009 due to the global recession, there is still an increasing trend. In the future, corresponding to a diversity of biodiesel fuels and to much severer environmental regulations, next-generation DPFs will be demanded.

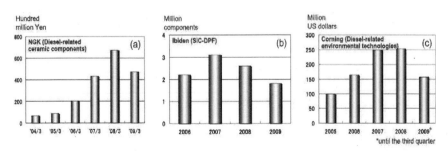

Figure 1. Current DPF-related production by major suppliers: (a) NGK, (b) Ibiden and (c) Corning. (The charts show the relative trends. Not for a comparison among manufacturers.)

In this proceedings paper, current DPF materials (first and second generations) are briefly introduced, and then, prospective third generation DPF materials are discussed. And then, our porous MgTi$_2$O$_5$, toward a candidate of third generation DPF material, will be demonstrated. Porous MgTi$_2$O$_5$ ceramic has been synthesized by one-step pyrolytic reacting sintering (*in situ* processing), where decomposed CO$_2$ gas from a carbonate source acts as an intrinsic pore forming agent.[5-9] Uniformly porous MgTi$_2$O$_5$ ceramic with very narrow pore size distribution has been successfully obtained. Some preliminary results on microstructure and pore structure are presented. Details of the porous MgTi$_2$O$_5$ and its composite (*viz.*, porous MgTi$_2$O$_5$/MgTiO$_3$) will be described in a forthcoming paper.[10]

FIRST, SECOND AND PROSPECTIVE THIRD GENERATION DPF MATERIALS

Cordierite (Mg$_2$Al$_4$Si$_5$O$_{18}$) and silicon carbide (SiC) are currently used for DPF materials; cordierite, the first generation, is characterized by its low-thermal expansion, low-cost and light-weight, while SiC, the second generation, is featured by its high-strength, high refractoriness, high thermal conductivity and so on (suitable for high-speed vehicles). Meanwhile, third generation DPF materials with low-thermal expansion, low-cost, high-strength and high refractoriness, are now under development. Up to now, several double oxides, such as mullite (3Al$_2$O$_3$·2SiO$_2$)[2,11], zircon (ZrSiO$_4$)[12] and aluminum titanate (Al$_2$TiO$_5$),[13-16] are of particular interest as candidates of the third generation, due to their low coefficients of thermal expansion (CTE) among oxides. Table I. summarizes an outline of DPF materials of first, second and prospective third generations.

Table I. Outline of first, second and prospective third generation DPF materials.

	First generation	Second generation	Third generation *Advanced and well-balanced*
Materials	Cordierite (Mg$_2$Al$_4$Si$_5$O$_{18}$)	Silicon carbide (SiC)	H.-T. Double oxides ?? *e.g.* Mullite ?? Aluminum Titanate ??
Cost merit	Excellent (Using natural sources)	Poor (Need atmosphere control)	Very good (Sinterable in air)
Refractoriness	So-so	Excellent	Very good (High melting points)
Chemical stability	Good	Good	Very good
Thermal shock resistance	Excellent	Good	Very good (Low thermal expansion)
Mechanical properties	So-so	Excellent	Very good
Notes	Applied for buses etc.	Applied for high-speed vehicles etc.	Will be applied for various (bio)diesel fuels under severe environmental regulations

Among several candidates for the third generation, Al$_2$TiO$_5$ (AT) has been eagerly studied due to its high thermal shock resistance caused by the low-thermal expansion.[17-20] Al$_2$TiO$_5$ has orthorhombic pseudobrookite-type crystal structure (generally expressed as Me$_3$O$_5$). US Corning and Japanese Ohcera Co./Sumitomo Chemical Co. are developing Al$_2$TiO$_5$-based DPF. Although Al$_2$TiO$_5$ has good thermal shock resistance and relatively high melting point, undoped Al$_2$TiO$_5$ tends to decompose into Al$_2$O$_3$ and TiO$_2$ at elevated temperatures (as shown in the next section). Furthermore, due to the strong anisotropy of thermal expansion, resultant microcracks gradually degrades mechanical properties. Hence, Al$_2$TiO$_5$-based composites or solid solutions (instead of pure Al$_2$TiO$_5$) have been actually developed for DPF applications. [17-20]

MgTi$_2$O$_5$ (MT$_2$) also has orthorhombic pseudobrookite-type crystal structure and also shows low thermal expansion. However the thermal-expansion anisotropy of MgTi$_2$O$_5$ is not so prominent as Al$_2$TiO$_5$.[18] MgTi$_2$O$_5$ is thermally more stable than Al$_2$TiO$_5$, and thus, MgTi$_2$O$_5$ is used as a "pseudobrookite-phase stabilizer" for Al$_2$TiO$_5$ phase to make Al$_2$TiO$_5$-MgTi$_2$O$_5$ solid solutions.[21-23] Actually, Kyocera Co. has proposed MgTi$_2$O$_5$-Al$_2$TiO$_5$ DPFs.[23] Although its prospective characteristics, MgTi$_2$O$_5$ is not so widely studied as Al$_2$TiO$_5$ (*e.g.*, 94 papers including "MgTi$_2$O$_5$" and 385 papers including "Al$_2$TiO$_5$" in Scopus database on February 2010) .

CERAMICS WITH PSEUDOBROOKITE-TYPE CRYSTAL STRUCTURE

Lattice constants and linear thermal expansion coefficients of pseudobrookite-type compounds Me$_3$O$_5$ reported by G. Bayer are shown in Table II.[18] Note that lattice constants in this table are shown as the *Cmcm*(63) space group, although *Bbmm*(63) notation is preferred in some recent reports. Al$_2$TiO$_5$ indicates negative thermal expansion in the direction of *a*-axis (*c*-axis for *Bbmm* notation). Due to its large thermal-expansion anisotropy, plenty of microcracks are formed in bulk forms, resulting in low thermal expansion. Despite this favorable low-thermal expansion characteristics, microcracks gradually deteriorate the fracture strength. Furthermore, due to the large distortion of MeO$_6$-octahedra, Al$_2$TiO$_5$ phase is metastable below 1200°C. Hence, under high-temperature conditions (< 1200°C), it gradually decomposes into Al$_2$O$_3$ and TiO$_2$. In order to overcome these weak points of Al$_2$TiO$_5$, Al$_2$TiO$_5$-based composite or solid solution are prepared. For example, Corning Co. have developed Al$_2$TiO$_5$-mullite-feldspar ([Sr/Ca]O· Al$_2$O$_3$· 2SiO$_2$) composite system for third-generation DPF.[13-16]

Besides Al$_2$TiO$_5$-based materials, other pseudobrookite-type compounds, with less distortion of MeO$_6$-octahedra, also become candidates for the third-generation DPF materials. Among various materials in Table II, MgTi$_2$O$_5$ does not show the negative thermal expansion behavior along with *a*-axis and has smaller thermal expansion along with *b*- and *c*-axes. Thus, it can exist as a thermally stable pseudobrookite-type compound without doping, different from Al$_2$TiO$_5$. Thinking about its low-cost, safe, and refractory composition, MgTi$_2$O$_5$ can be another candidate for the third-generation DPF material.

Table II. Lattice constants and linear thermal expansion coefficients of pseudobrookite-type compounds Me$_3$O$_5$ by G. Bayer.[18] Note that lattice constants and linear thermal expansion coefficients are shown as *Cmcm*(63) space group.

Compound	Lattice constants (Å) at			Linear thermal expansion (×10^{-6}/°C)	
	20°C	520°C	1020°C	20-520°C	20-1020°C
Al$_2$TiO$_5$	a = 3.5875	3.5823	3.5768	β_a = −2.9 ± 0.2	−3.0 ± 0.3
	b = 9.4237	9.4721	9.5339	β_b = 10.3 ± 0.6	11.8 ± 0.6
	c = 9.6291	9.7262	9.8393	β_c = 20.1 ± 1.0	21.8 ± 1.1
Ga$_2$TiO$_5$	a = 3.6174	3.6127	decomposes	β_a = −2.6 ± 0.2	
	b = 9.8196	9.8613		β_b = 8.5 ± 0.5	
	c = 10.0054	10.1020		β_c = 19.3 ± 1.0	
Fe$_2$TiO$_5$	a = 3.7385	3.7399	3.7406	β_a = 0.7 ± 0.1	0.6 ± 0.1
	b = 9.7954	9.8434	9.8947	β_b = 9.8 ± 0.6	10.1 ± 0.6
	c = 9.9853	10.0577	10.1485	β_c = 14.6 ± 0.7	16.3 ± 0.8
MgTi$_2$O$_5$	a = 3.7442	3.7486	3.7529	β_a = 2.3 ± 0.2	2.3 ± 0.2
	b = 9.7363	9.7755	9.8418	β_b = 8.1 ± 0.4	10.8 ± 0.5
	c = 9.9870	10.0526	10.1450	β_c = 13.2 ± 0.7	15.9 ± 0.8
MgTi$_{1.6}$Zr$_{0.4}$O$_5$	a = 3.7646	3.7685	3.7740	β_a = 2.1 ± 0.1	2.5 ± 0.2
	b = 9.8159	9.8698	3.9202	β_b = 11.0 ± 0.6	10.6 ± 0.7
	c = 10.0458	10.1175	10.2078	β_c = 14.3 ± 0.7	16.2 ± 0.6
MgFeNbO$_5$	a = 3.7771	3.7810	decomposes	β_a = 2.1 ± 0.2	
	b = 9.9858	10.0174		β_b = 6.4 ± 0.5	
	c = 10.1868	10.2589		β_c = 14.1 ± 0.7	
MgFeTaO$_5$	a = 3.7879	3.7901	3.7953	β_a = 1.2 ± 0.2	1.9 ± 0.2
	b = 9.9988	10.0373	10.0896	β_b = 7.8 ± 0.5	9.1 ± 0.6
	c = 10.1790	10.2396	10.3185	β_c = 12.1 ± 0.7	13.7 ± 0.9

Phase diagrams of Al$_2$O$_3$-TiO$_2$ and MgO-TiO$_2$ systems are demonstrated in Figure 2.[24,25] As mentioned above, Al$_2$TiO$_5$ is a metastable phase below 1200°C, while MgTi$_2$O$_5$ is a stable phase. It is reported that even after the heat-treatment at 700°C for 162 h, crystal structure of the MgTi$_2$O$_5$ remained as pseudobrookite.[26]

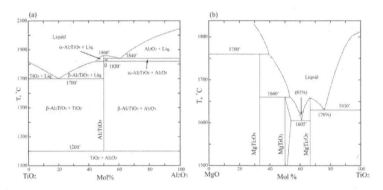

Figure 2. Phase diagram of (a) Al$_2$O$_3$-TiO$_2$ and (b) MgO-TiO$_2$.[24,25]

POROUS $MgTi_2O_5$ BY REACTIVE SINTERING

In this section, porous $MgTi_2O_5$ prepared by reactive sintering method is demonstrated. Similarly to previously reported UPC-3D[5-9] (uniformly porous ceramic/composite with three dimension network structure), magnesium carbonate is used as a starting materials; decomposed CO_2 gas, emitted during the one-step pyrolytic reacting sintering, forms very uniform open porous structure.

Commercially available $MgCO_3$ and TiO_2 anatase (99.9% purity, Kojundo Chemical Laboratory Co. Ltd., Saitama, Japan), and LiF (99.9%, Wako Pure Chemical Ind., Ltd, Osaka, Japan) were used as the starting powders. LiF acts as a mineralizer. Note that low-cost natural resources may be replaceable for magnesium carbonate and TiO_2. It is an advantage of $MgO-TiO_2$ system.

The $MgCO_3$ and TiO_2 powders (Mg:Ti = 1:2 in mole) with LiF (0.5 mass% for total starting powders) were wet-ball milled in ethanol for 2 h in a planetary ball-mill (acceleration: 4 G). The mixed slurry was dried and then sieved through a 150-mesh screen. To obtain bulk porous $MgTi_2O_5$, the mixed powder was cold isostatically pressed at 200 MPa after mold-pressing. The green compacts with no binder, 15 mm in diameter and ~ 3 mm in thickness (cylinder), were sintered in air at 1000-1300°C for 2 h to obtain the porous $MgTi_2O_5$. The microstructure of porous $MgTi_2O_5$ was characterized using a scanning electron microscope (SEM, JSM-6500F, JEOL, Tokyo, Japan). The pore-size distribution was determined by the mercury porosimetry. The reaction behavior of the mixed powder and constituent phases of the bulk materials were analyzed by powder X-ray diffraction. (XRD, RIGAKU Multiflex, Cu-Ka, 40 kV and 40 mA). XRD analysis revealed that the obtained porous ceramic was composed of $MgTi_2O_5$ (the detail will be shown in a forthcoming paper[10]).

Figure 3 shows microstructure and pore-size distribution of porous $MgTi_2O_5$ sintered at 1100°C. Corresponding to the anisotropic crystal structure, elongated $MgTi_2O_5$ grains were formed during the *in situ* processing. The porous $MgTi_2O_5$ (porosity: 42%) had very narrow pore-size distribution with a peak size at 1.0 μm, in good agreement with the SEM observation. Although the pore-size of 1.0 μm is rather small for ordinary DPF application (~10 μm), it can be further controlled *e.g.*, by the addition of sacrificial pore-forming agent (like porous cordierite) or by the use of coarse grains (like porous SiC). Besides, the pore-size of 1.0 μm might be suitable for much severer filtration purposes (*e.g.*, PM2.5 or smaller). Considering about the very narrow pore-size distribution, this material can be also used as a standard material for various measurement.

SUMMARY AND REMARKS

In this proceedings paper, current DPF materials and prospective third-generation DPF materials are discussed. Among pseudobrookite-type compounds, $MgTi_2O_5$ is a potential candidate for the third-generation DPF material. Porous $MgTi_2O_5$ prepared by reactive sintering method was briefly introduced. The porous $MgTi_2O_5$ sintered at 1100°C exhibited very narrow pore-size distribution at 1.0 μm. Thinking about the *in situ* reaction relating to CO_2, the porous $MgTi_2O_5$ might absorb CO_2 gas at high temperatures during DPF-regeneration. Intelligent DPF with "self-healing function" for microcracks might be produced with this material. Details of the porous MgTi$_2$O$_5$ and porous MgTi$_2$O$_5$/MgTiO$_3$ composite will be described in a forthcoming paper.[10]

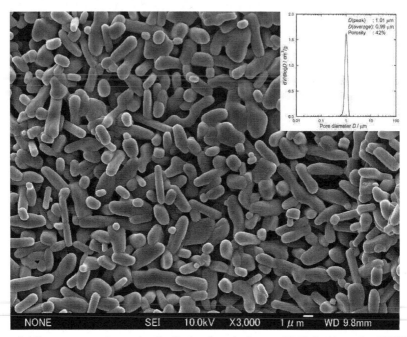

Figure 3. Microstructure and pore-size distribution (insert) of porous MgTi$_2$O$_5$ (sintered at 1100°C for 2h) with pseudobrookite-type crystal structure. Elongated grains and very narrow pore-size distribution at 1.0 μm were confirmed.

ACKNOWLEDGMENT

This work was supported by Grant-in-Aid for Science Research No. 19685020 For Young Scientist: Category A. The author thanks to Ms. Machiko Matsumura for the help of industrial survey.

REFERENCES

[1]J. Adler, Ceramic Diesel Particulate Filters, *Int. J. Appl. Ceram. Tech.*, **2** [6], 429-439 (2005).

[2]A. J. Pyzik and C. G. Li, New Design of a Ceramic Filter for Diesel Emission Control Applications, *Int. J. Appl. Ceram. Tech.*, **2** [6], 440-451 (2005).

[3]C. K. Narula, C. Stuart Daw, J. W. Hoard and T. Hammer, Materials Issues Related to Catalysts for Treatment of Diesel Exhaust, *Int. J. Appl. Ceram. Tech.*, **2** [6], 452-466 (2005).

[4]K. Ohno, The SiC Porous Material Technology which Actualizes the Development to Diesel Engine Automobile Business, (in Jpn.) *Ceram. Jpn.*, **42** [6] 431-438 (2007).

[5]Y. Suzuki, P. E. D. Morgan and T. Ohji, New Uniformly Porous CaZrO₃/MgO Composites with Three-Dimensional Network Structure from Natural Dolomite, *J. Am. Ceram. Soc.*, **83** [8] 2091-93 (2000).

[6]Y. Suzuki, N. Kondo and T. Ohji, In Situ Synthesis and Microstructure of Porous CaAl₄O₇/CaZrO₃ Composite, *J. Ceram. Soc. Jpn.*, **109** [3] 205-209 (2001).

[7]Y. Suzuki, N. Kondo, T. Ohji and P. E. D. Morgan, Uniformly Porous Composites with 3-D Network Structure (UPC-3D) for High-Temperature Filter Applications, *Int. J. Appl. Ceram. Tech.*, **1** [1] 76-85 (2004).

[8]Y. Suzuki, M. Tsukatsune, S. Yoshikawa, and P. E. D. Morgan, Uniformly Porous Al₂O₃/LaPO₄ and Al₂O₃/CePO₄ Composites with Narrow Pore-Size Distribution, *J. Am. Ceram. Soc.*, **88** [11] 3283-3286 (2005).

[9]Y. Suzuki and P. E. D. Morgan, Meso- and Macroporous Ceramics by Phase Separation and Reactive Sintering Methods, *MRS Bull.*, **34** [8] 587-591 (2009).

[10]Y. Suzuki, *In Situ* Processing of Uniformly Porous MgTi₂O₅ and MgTi₂O₅/MgTiO₃ Toward Third Generation Diesel Particulate Filter Materials, *in preparation*.

[11]A. J. Pyzik, C. S. Todd and C. Han, Formation Mechanism and Microstructure Development in Acicular Mullite Ceramics Fabricated by Controlled Decomposition of Fluorotopaz, *J. Eur. Ceram. Soc.*, **28** [2] 383-391 (2008).

[12]G. Del Pin, S. Maschio, S. Brückner and A. Bachiorrini, Thermal Interaction between Some Oxides and Zircon as a Material for Diesel Engines Filter, *Ceram. Int.*, **30** [2] 279-283 (2004).

[13]S. B. Ogunwumi, P. D. Tepesch, T. Chapman, C. J. Warren, I. M. Melscoet-Chauvel and D. L. Tennent, Aluminum Titanate Compositions for Diesel Particulate Filters, *SAE Technical Paper*, 2005-01-0583 (2005).

[14]R. S. Ingram-Ogunwumi, Q. Dong, T. A. Murrin, R. Y. Bhargava, J. L. Warkins and A. K. Heibel, Performance Evaluations of Aluminum Titanate Diesel Particulate Filters, *SAE Technical Paper*, 2007-01-0656 (2007).

[15]W. A. Cutler and T. Boger, A. F. Chiffey, P. R. Phillips, D. Swallow and M. V. Twigg, Performance Aspects of New Catalyzed Diesel Soot Filters Based on Advanced Oxide Filter Materials, *SAE Technical Paper*, 2007-01-1268 (2007).

[16]D. Rose, O. A. Pittner, C. Jaskula and T. Boger, T. Glasson and V. Miranda Da Costa, On Road Durability and Field Experience Obtained with an Aluminum Titanate Diesel Particulate Filter, *SAE Technical Paper*, 2007-01-1269 (2007).

[17]B. Morosin and R. W. Lynch, Structure Studies on Al₂TiO₅ at Room Temperature and at 600°C, *Acta Cryst., B*, **28**, 1040-1046 (1972).

[18]G. Bayer, Thermal Expansion Characteristics and Stability of Pseudobrookite-type Compounds,

Me$_3$O$_5$, *J. Less-Common Metals*, **24** [2] 129-138 (1971).

[19]S. T. Norberg, N. Ishizawa, S. Hoffmann, and M. Yoshimura, Redetermination of β-Al$_2$TiO$_5$ Obtained by Melt Casting, *Acta Cryst. E*, **61** [8] i160-i162 (2005).

[20]I. J. Kim and L. J. Gauckler, Excellent Thermal Shock Resistant Materials with Low Thermal Expansion Coefficients, *J. Ceram. Proc. Res.*, 9 [3] 240-245 (2008).

[21]T. Shimazu, H. Maeda, E. H. Ishida, M. Miura, N. Isu, A. Ichikawa, K. Ota, High-Damping and High-Rigidity Composites of Al$_2$TiO$_5$-MgTi$_2$O$_5$ Ceramics and Acrylic Resin, *J. Mater. Sci.*, **44** [1] 93-101 (2009).

[22]T. Shimazu, M. Miura, N. Isu, T. Ogawa, K. Ota, H. Maeda, E. H. Ishida, Plastic Deformation of Ductile Ceramics in the Al$_2$TiO$_5$-MgTi$_2$O$_5$ system, *Mater. Sci. Eng. A*, **487** [1-2] 340-346 (2008).

[23]Kyocera Co., *International patent application*, PCT/JP2008/067481, WO/2009/041611.

[24]P. Pena, S. DeAza, The System Zirconium Dioxide-Aluminum Oxide-Titanium Dioxide, *Ceramica (Florence)*, **33** [3] 23-30 (1980). (NIST-ACerS Phase Equilibria Diagrams Database No. 92-008.)

[25]I. Shindo, Determination of the Phase Diagrams by the Slow Cooling Float Zone Method: the System MgO-TiO$_2$, *J. Cryst. Growth*, **50** [4] 839-851 (1980). (NIST-ACerS Phase Equilibria Diagrams Database No. 92-003.)

[26]JCPDS-ICDD Powder Diffraction File, MgTi$_2$O$_5$, no. 35-0796 (1985).

ALUMINUM TITANATE COMPOSITES FOR DIESEL PARTICULATE FILTER APPLICATIONS

Monika Backhaus-Ricoult, Chris Glose, Patrick Tepesch, Bryan Wheaton*, Jim Zimmermann
Crystalline Materials Research or Characterization Science*, Science & Technology
Corning Incorporated, Corning NY 14831, USA

ABSTRACT

Porous composites of aluminum titanate and mixed strontium/calcium feldspar are used as a material in diesel particulate filter applications because of their low thermal expansion, high thermal shock resistance, high heat capacity, high chemical resistance to ash and their ability to be formed into porous materials with porosity >50% with pore sizes that allow efficient filtering of particulate matter within a wide size range without creating an undesirably high backpressure. In the present work, reaction-templated growth of aluminum titanate is explored, where during its high temperature formation aluminum titanate conserves the template morphology and develops texture. The growth of aluminum titanate from simple oxide starting materials is constrained by a number of thermodynamic, kinetic, crystallographic and morphological factors. We have investigated the impact of template shape and crystallographic orientation in single crystalline and polycrystalline reacting mixtures on the material microstructure, its crystallographic texture and the resulting material porosity and thermo-mechanical properties. Reaction-templating is used to derive materials that cover a wide range of porosity, pore size and shape and, in specific cases, also demonstrate enhanced strength.

INTRODUCTION

Limited raw oil resources, increasing fuel cost, concerns around carbon footprint and CO_2 emissions, together with the development of improved diesel engine technologies are the key drivers for the growing market of diesel engines for passenger cars and trucks. Since small soot particles constitute a severe health issue, the particulate emission of diesel engines is subject to strong legislative regulation. Low particulate emissions and removal of the most dangerous fine particles can be achieved by use of diesel particulate filters.

Diesel particulate filters of different designs and materials are available. Corning has developed highly efficient wall-through-flow filters of cordierite and aluminum titanate-based composites[1]. SiC is another material currently used to make diesel particulate filters. Cordierite and Corning's aluminum titanate based DuraTrap® AT material both exhibit very low thermal expansion and high thermal shock resistance[2,3]. While application of cordierite is limited by its melting temperature (1450°C) and lower heat capacity, Duratrap® AT excels by the much higher melting point of aluminum titanate and a superior heat capacity of the composite[2]. It also demonstrates a superior thermal shock resistance to silicon carbide. SiC has a higher intrinsic heat capacity and heat conductivity, so that local temperature spikes should be reduced by a higher heat flow; however, due to its higher thermal expansion, its thermal shock resistance remains low (making segmentation necessary for its use in filters).

In the past, formation of aluminum titanate from titania and alumina has been extensively studied[4,5,6,7], revealing that, due to the small thermodynamic driving force at low temperatures, its nucleation is difficult, occurs mainly at impurities or glass pockets and remains rate controlling (temperature range from the thermodynamic stability limit of about 1280°C to approximately 1350°C), while for higher temperatures the reaction becomes diffusion-limited. Based on the phase distribution in the reacting powder mixture, a fast and a slow diffusion-controlled growth mode were observed[7]. Al_2TiO_5 formation was observed to be fast in an interconnected titania surrounding due to the high solubility and high mobility of alumina in TiO_2. Diffusion-controlled growth was reported to be several orders of magnitude slower once a continuous aluminum titanate network had formed and diffusion had to go through the aluminum titanate product layer. Crystallographic aspects of the solid state reactions have not been fully investigated.

Texture and microstructure engineering through reaction-templating has been successfully used for many ceramics to improve properties of interest[8]. In the present work, high temperature reaction-templating is used for making aluminum titanate-feldspar composites with the aim to impose morphology and texture through a substrate. The main focus of the study is on the impact of template shape and crystallographic orientation in reacting powder mixtures on the material microstructure, its crystallographic texture and the resulting porosity and thermo-mechanical properties.

EXPERIMENTAL

Material Processing

Alumina, titania, silica, strontium carbonate, and calcium carbonate were mixed with sinter additives, pore former and binder and either cold pressed and fired or extruded into a honeycomb, dried and then fired. Standard commercial powders were used. In one part of this study, titania powder was substituted by 1mm sized pieces of rutile single crystals with either (100) or (010) polished surface termination. In another part of the study, the alumina powder was substituted by either 1mm sized pieces of corundum single crystals with (0001) polished surface termination, small alumina rods with an approximate diameter of 3 micrometers and length ranging within 50 and 150 micrometers or alumina platelets of average thickness 1-2 micrometers and diameters of 20-40 μm. Powder mixtures with single crystal pieces were cold-pressed, the others extruded into honeycomb with about 300 cells per square inch and wall thickness of 13 mils, referred to as (300/13). All materials were fired in a box furnace in air up to various top temperatures for different durations of time. The reaction of powder mixtures was considered complete when no further formation of aluminum titanate occurred.

Material Characterization

Phase analysis of the different materials was conducted by XRD with Rietveld refinement utilizing a PANalytical MPD equipped with an X'Celerator multiple strip detector. The distribution of porosity and phases at the microstructural level was visualized by SEM on polished sample cross sections and quantified with help of image analysis. For an evaluation of the anisotropy of the porosity, sections parallel and perpendicular to the extrusion direction were analyzed so that an average elongation of the porosity could be derived from the pore extensions in both directions.

Electron backscattered diffraction on the SEM was used for orientation mapping of polished sample sections to derive grain size, relative orientation and texture of the phases in presence with respect to the honeycomb extrusion direction or a sample pressing direction. Appropriate standards and commercial software allowed attribution of each electron backscatter pattern with its intersecting Kikuchi lines to phase and crystallographic orientation. EBSD analysis was realized on a Hitachi SU70 SEM, equipped with an Oxford/HKL EBSD system. The SEM was typically operated at 20kV and a beam current of about 20nA. The beam was stepped with 0.2 to 2μm step size over a representative area of about 1000 microns x 700 microns. Phases used for identification of fired materials included aluminium titanate, corundum, rutile and feldspar. While collecting the EBSD data, energy dispersive spectroscopy was used to simultaneously collect chemical information. Post-collection analysis of the EBSD data included noise reduction by removing wild spikes and smoothing. Pole figures were generated using HKL Mambo software with 5° data clustering.

In-situ SEM analysis was realized on a FEI Quanta 200 environmental field emission scanning electron microscope equipped with a programmable heating stage. As-fired sample surfaces of aluminium titanate – feldspar composites were observed in-situ at high temperature. Samples were heated at a rate of 10°C/min up to 1200°C, held a 1200°C for 10 minutes followed by controlled cooling to room temperature. Images were collected throughout the heating and cooling cycle. Microcracks and their evolution were tracked.

Material Property Evaluation

Thermal expansion was measured for bar-shaped samples with dimension 0.25" x 0.25"x 2" during heating from room temperature to 1200°C and subsequent cooling to room temperature. MOR was measured by four point flexure with a lower span of 2" (50.8mm) and an upper span of 0.75" (19mm). The specimen geometry for the 4-point flexure tests was 2.5" (63.5mm) in length, 0.5" (12.7mm) in width and 0.25" (6.4mm) thick. The force-measuring system used was equipped with a read-out of the maximum force and a calibrated load cell. The MOR value was calculated using the well known flexure strength equation $\sigma_{4-po\,\text{int},\,MOR} = 3/4\,PL/bd^2$, however, this equation does not account for the cellular channels through the specimen and is not the true strength of the material. All specimens tested had a square cellular (honeycomb) geometry with the channels in the direction of the length. The actual material strength, often referred to as the wall strength, was determined by accounting for the cellular structure (ASTM standard C1674-08).

Bar-shaped samples with dimension 5" x 1" x 0.5" and the long axis being oriented in the direction of the honeycomb channels were used to measure the elastic modulus by flexural resonance frequency. Samples were heated to 1200°C or 1400°C and cooled back to room temperature. For each temperature the elastic modulus was directly derived from the resonance frequency and normalized for sample geometry and weight by referring to ASTM C 1198-01.

RESULTS

Reaction-Templated Aluminum Titanate Growth On Single Crystals Of Rutile Or Corundum

Reacted mixtures of the raw materials were analyzed after exposure to various firing temperatures and durations. Figure 1 illustrates the microstructure of a rutile single crystal after reaction with the powder mixture. A mixture of aluminum titanate and feldspar formed on the single crystal surface. No continuous reaction layer was observed, but formation of individual aluminum titanate crystals. After 10h at 1400°C, the aluminum titanate crystallite size reached about 7 μm. The individual crystals showed distinct orientations. Electron backscattered diffraction of the rutile single crystal (see Figure 1) and the aluminum titanate crystals (see Figure 1) did not reveal any particular orientation relationship between the two phases. However, a clear correlation was established between the surface plane of the single crystal and the growing aluminum titanate crystals. The a-axis of aluminum titanate was aligned perpendicular to the underlying rutile surfaces and the c-axis of aluminum titanate was oriented within the substrate plane. The same preferential growth direction of aluminum titanate was observed for many different reaction temperatures.

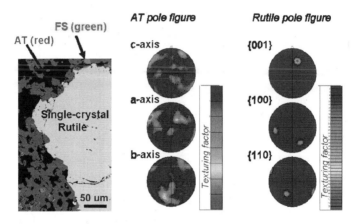

Figure 1: EBSD phase map showing aluminum titanate growth from the oxide mixture on a rutile single crystal together with the corresponding pole figures of rutile and alumina titanate.

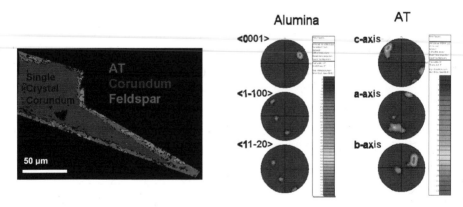

Figure 2: EBSD phase map showing aluminum titanate growth from the oxide mixture on a corundum single crystal together with pole figures of alumina and alumina titanate

The same features were found for aluminum titanate grown from an oxide mixture on corundum single crystals of various orientations, Figure 2. The phase image in Figure 2 was acquired by electron backscattered diffraction imaging and shows the corundum single crystal with its reaction layer of aluminum titanate (red) and feldspar (green). Again the pole figures of alumina and aluminum titanate do not indicate any particular orientation relationship. However, the three families of high intensity in the partial pole figure of the aluminum titanate a-axis each correspond to one of the directions perpendicular to the single crystal surfaces, so that we can conclude that aluminum titanate grows on an alumina substrate plane with its a-axis pointing perpendicular to the substrate and the c-axis lying in the substrate plane.

Templated Growth Of Aluminum Titanate On Alumina Particles In Reacting Oxide Powder Mixtures

With the goal to improve the texturing statistics from our single crystal study and evaluate the impact of the reaction-templating on the microstructure and properties of porous aluminum titanate-feldspar composites, we varied the morphology of the reacting alumina powder in the oxide powder mixture, using roughly spherical, rod-shaped and platy particle shapes of alumina. Figures 3 a, b, c present typical particle shapes. The round, rod-shaped and platy alumina particles that we used had similar particle volume, but very different aspect ratios. The resulting microstructures of the fired materials differed; Figures 3 d, e, f present as-fired surfaces. Since these materials were obtained by extrusion, particles with anisotropic shape (rods and plates) preferentially aligned during the extrusion with their long axis in the extrusion direction. The reference microstructure obtained from the round alumina particles demonstrated a rather isotropic pore shape and regular matter distribution, Figure 3f. As-fired surfaces of the material made from platy alumina, Figure 3d, exhibited a platy grain shape, a small anisotropy in pore shape and an irregular distribution of porosity (see polished cross sections in Figure 3g). The material derived from rod-shaped alumina showed even larger anisotropy. The fired surface, Figure 3e, had preserved the characteristics of the alumina rod-shaped particles and also reflected a preferential alignment as result of the extrusion. That alignment produced an anisotropic pore structure, Figure 3h, with average pore dimensions being rather isotropic in the plane perpendicular to the extrusion direction and showing some elongation in the extrusion direction. We determined an elongation factor of 1.4 for the anisotropy of the porosity, compared to 1 for the material made from isotropic alumina particles.

EBSD orientation mapping was conducted on the material sections parallel and perpendicular to the extrusion direction. Aluminum titanate pole figures for the sections parallel to the extrusion direction are shown in Figures 3j, k, l. The material obtained from the round particles shows a random orientation of its aluminum titanate grains; the lack of texture is visible in Fig. 3l by the homogeneity of intensity in the partial pole figures for all main directions. Materials made from alumina platelets show texture for aluminum titanate, Figure 3j, with a preferential alignment of the a- and b-axis perpendicular to the extrusion direction and a weak preferential alignment of the aluminum titanate c-axis in the direction of the extrusion (vertical). The texture remained weak with texture factors smaller than 2.5 for the a-axis and smaller than 2 for the c-axis. Use of rod-shaped alumina particles produced a stronger texture, Figure 3k. The texturing factor of the aluminum titanate crystallographic c-axis reached 4, demonstrating a strong preferential alignment of the c-axis in the direction of extrusion (vertical), while a- and b-axis are quite homogeneously distributed in the plane perpendicular to the extrusion (the entire plane has an increased intensity with a texture factor of about 2).

The observed crystallographic texture can again be explained by the reaction-templated growth mechanism that was discussed for the alumina single crystalline template. Aluminum titanate grows on the alumina grain substrate with its a-axis perpendicular to the substrate surface and its c-axis in the substrate plane. Since anisotropic alumina particles in the powder mixtures align upon extrusion with plate and rod long axis preferentially pointing into the extrusion direction, the reaction-templated growth of aluminum titanate produces an overall texture in the material that directly results from the relative surface sizes and the orientation of the alumina particle in the powder mixture prior to the reaction.

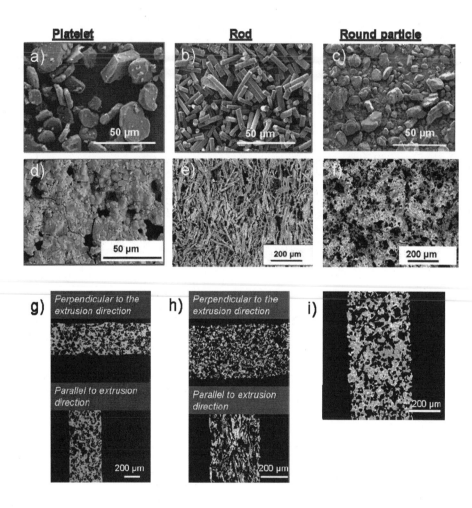

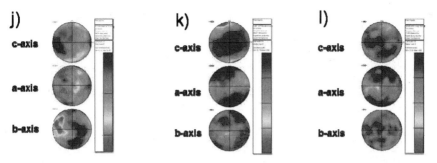

Figure 3 - Representative images of platy (a), rod-shaped (b) and round (c) alumina particles, fired surfaces of materials obtained from platy (d), rod-shaped (e) and round (f) alumina, (g,h,i) polished cross sections parallel and perpendicular to the extrusion direction (vertical direction), (j,k,l) aluminum titanate pole figures for c, a, and b axis of fired material cross sections parallel to the extrusion direction with the extrusion direction pointing vertical in the image plane.

Thermal Expansion Anisotropy in Aluminum Titanate and Microcracking in Polycrystalline Ceramics

Figure 4 displays the changes of aluminum titanate lattice parameters with temperature from room temperature to 1400°C as determined by high temperature X-ray diffraction[9]. Due to the anisotropy in thermal expansion, negative expansion c-axis and positive expansion a and b axes, significant stresses build up upon heating and cooling aluminium titanate-based ceramics. Such stresses can be released by microcracking[10]. Figure 5 displays an SEM micrograph of a typical microcracked aluminium titanate grain in a composite material. The EBSD-generated pole figure of the corresponding grain reveals the crystallographic plane of the microcracks, Figure 5; the microcracks disconnect the aluminium titanate grain perpendicular to its highest expansion b-axis and extend parallel to the low expansion c-axis.

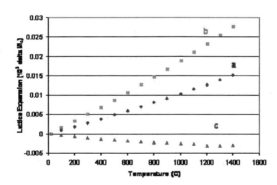

Figure 4: Lattice expansion as a function of temperature for aluminum titanate.

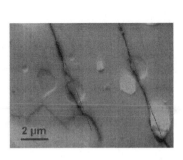

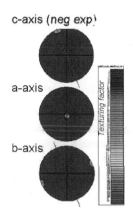

Figure 5: SEM micrograph and corresponding EBSD pole figure of a representative microcracked aluminum titanate in the aluminum titanate-feldspar composite material. The red lines in the SEM image indicate the global microcrack plane; their orientation is also indicated by a red line in the partial pole figure.

Reversible Formation and Healing of Microcracks in Aluminum Titanate-Feldspar composites

We have observed surfaces of AT-feldspar composites in situ in the SEM during heating up to 1200°C and cooling back to room temperature, Figure 6. In the initial material, at room temperature, microcracks of different sizes were found. Upon heating, the narrower microcracks in smaller grains closed at temperatures below 1000°C, see top row of images in Figure 6, while broader microcracks in larger grain areas require higher temperature, see bottom row images in Figure 6. From the experimental top temperature of 1200°C, the material was cooled back to room temperature. Cracks in larger grains were observed to open earlier (bottom row of Figure 6) than the microcracks in smaller grains. A comparison of the precise locations of the microcracks at the start of the heating cycle and after cooling showed that the microcracks globally re-adopted their original crack plane, following crack propagation statistics. However, differences in the precise crack path were found that indicate that the microcracks had not only closed, but healed at high temperature. Healing of microcracks and microcrack reformation upon cooling is also supported by the hysteresis observed in heating-cooling cycles during in-situ SEM or Young's modulus measurements, Figure 7.

A rough analysis of images prior to heating and after cooling indicated that the microcrack density in the material remained roughly the same upon several cycles. A more precise analysis of the microcrack density and its evolution upon temperature cycling was derived from measurements of the elastic modulus during temperature cycling.

Microcrack densities in the AT-based composites were derived from the high temperature elastic modulus and calculated as a function of temperature using the well known microcrack density – Young's modulus relationship (Equation 1) initially proposed by Walsh[11]. Elastic modulus (E) was measured during heating to 1400°C and cooling back to room temperature using the sonic resonance method. The temperature dependence of the non-microcracked

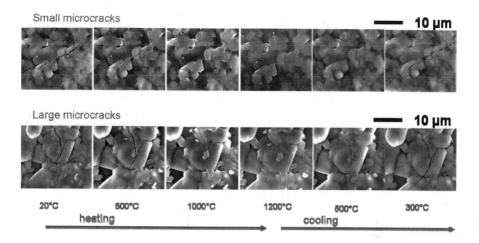

Figure 6: SEM images of AT-feldspar composite surfaces as observed in-situ during heating and cooling. The upper row of images shows a fine micorcrack within an assembly of small grains, the bottom row illustrates the behavior of a set of large microcracks in large grains. Image labeling indicates temperature, as well as cooling or heating part of the thermal cycle.

Young's modulus (E_0) was calculated using the Wachtman Equation[12], which was modified for a porous body within a honeycomb structure (Equation 2). Honeycomb geometry (open frontal area V_{OFA}) and porosity (P) were taken into account. The room temperature Young's modulus of a dense microcrack free material (E_0') was estimated to be 250GPa and the Poisson's ration (v) of the microcracked body was estimated to be 0.2 and the empirical fitting parameter ($T0$) estimated to be 375°C. The non-microcracked regions of the Young's modulus curves were fitted to the highest temperatures (fitted curves are incorporated as solid lines in Fig. 7).

$$\frac{E}{E_0} = \left(1 + \frac{16\left(1 - v^2\right)Nl^3}{9\left(1 - 2v\right)}\right)^{-1}$$

Equation 1

$$E_0 = E_0'\left(1 - V_{OFA}\right)\left(1 - P\right)^2 - b\,T\,e^{-T_0/T}$$

Equation 2

Figure 7 shows the elastic modulus evolution for three different composites during heating from room temperature to 1400°C and subsequent cooling back to room temperature. All materials exhibit a strong hysteresis with closure of the hysteresis loop around 1200°C. The absolute values of the material's elastic modulus differ due to differences in porosity and microcrack density. The room temperature microcrack densities (Nl^3) of the considered composite materials range from 3 to 7. The microcrack density curves converge for all materials to zero at approximately 1150°C during heating (Figure 7). The average microcrack density in the microcracked materials is stable, and the engineered microcracks can be used to engineer stable, desirable properties such as low CTE in a material.

It shall be underlined that there is an important difference between engineered microcracked materials like cordierite and aluminum titanate composites, which are stable over many hundred thermal cycles, and materials that suffer formation of macrocracks. Macrocracks do not heal upon heating; they extend during thermal cycling and lead to crack-failure after repeated thermocycling.

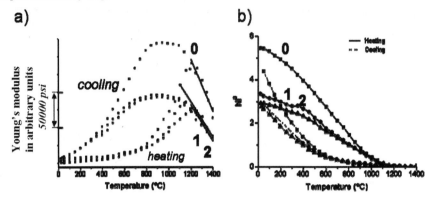

Figure 7: Aluminum titanate-feldspar research composites of different porosity, texture, aluminum titanate grain size and shape (labeled as research materials 0, 1, 2); a) elastic modulus as a function of temperature during heating and cooling cycle with the elastic modulus being indicated in relative units with the scale bar being 50000 psi, b) microcrack densities NI^3 as function of temperature during heating (solid lines) and cooling (dotted lines).

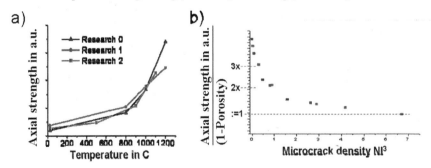

Figure 8: a) MOR of aluminum titanate - feldspar composites with different initial microcrack densities as function of temperature, b) specific MOR (MOR/(1-porosity)) in arbitrary units as a function of microcrack density NI^3. Arbitrary units are referenced to the base line of our most microcracked material at room temperature (dotted line with value = 1), so that the arbitrary

strength is indicated in reference to the room temperature strength of a research material 0 (blue) made from round alumina.

The strength of the microcracked composites relates directly to the microcrack density. Figure 8a shows the 4 point flexure MOR of composites with very different initial microcrack densities as function of temperature, while Figure 8b presents the normalized specific MOR (MOR/(1-porosity)) of these materials as a function of microcrack density NI^3. The outermost right point in Figure 8 for each material is its room temperature specific strength. The specific strength of the composites increases with temperatures as the microcracks close. The materials of this study with a wide range of room temperature microcrack densities and microstructure all fit a common universal curve.

Comparison Of The Thermomechanical Properties Of Materials Made From Round, Platy and Rod-Shaped Alumina

The thermomechanical properties of (300/13) honeycomb made from round, platy or rod-shaped alumina as presented in Figure 3 were measured.

Differences in the axial and radial thermal expansion behavior were observed. Figure 9 illustrates the differences for materials with the same porosity that were made from round alumina and rod-shaped alumina. The axial thermal expansion curves of the material derived from rod-shaped alumina shows a lower thermal expansion in the axial direction of the honeycomb and a higher thermal expansion in the radial honeycomb direction due to the preferential crystallographic alignment of the aluminum titanate negative expansion c-axis in the material derived by reaction templating and extrusion from rod-shaped alumina particles. This difference was less pronounced for materials made from alumina plates.

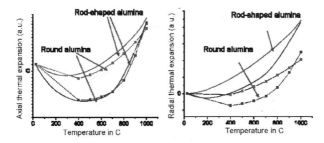

Figure 9: Comparison of axial (left side graph) and axial (right side graph) thermal expansion during heating and cooling for aluminum titanate –feldspar composites made from round alumina (continuous line) and rod-shaped alumina particles (line with square symbols); heating curves are shown in red, cooling curves are shown in blue.

The elastic modulus was measured as function of temperature for honeycomb materials covering a porosity range from 40 to 70% and being made either with rod-shaped, platy or round alumina. Figure 10 summarizes room temperature and 1000°C (heating curve) data for round and rod-shape alumina derived aluminum titanate composites. The elastic modulus data of all materials made from the round alumina particles are assembled on one straight line for room temperature and one line for 1000°C in the log E-mod – log (1-porosity) plot. The materials derived from rod-shaped alumina are also aligned on such two lines with the same slope, but high absolute elastic modulus values. At same porosity, the elastic modulus of materials made from rod-shaped alumina is higher than that of the corresponding material made from round alumina. The difference is larger at room temperature than at 1000°C and vanishes with further increasing temperature, as can be easily explained by the closure of the microcracks. Figure 10 presents relative values of the elastic modulus that are valid for dense materials and all honeycomb geometries. The elastic modulus of the 50% porosity material made from round alumina is fixed as reference point 100%.

The strength of extruded honeycomb materials with (300/13) cell geometry made from round and rod-shaped alumina particles was measured for materials covering a wide range of material porosity by 4 point flexure testing. Room temperature MOR results for the honeycomb axial and radial direction are shown in Figure 11, which presents a relative MOR in comparison to a 50% porosity material made from round alumina particles. At same porosity, axial and radial MOR were significantly higher for the materials derived from rod-shaped alumina particles. In the MOR – porosity plots of Figure 11, the data points were fitted by representative behavior for the materials derived from round and rod-shaped alumina particles. It shall be underlined that similar axial MOR values were achieved for a rod-shaped particle-derived material with 67% porosity and about 50% porosity materials made from the round alumina particles of this study.

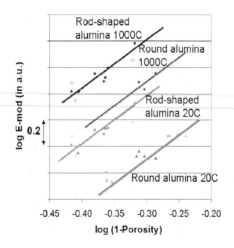

Figure 10 : Comparison of the room temperature (RT) and 1000°C elastic modulus as function of porosity for aluminum titanate –feldspar composites made from round alumina (green) and rod-shaped alumina (blue). Relative values are shown in comparison to a 50% porosity material made from round alumina particles; all measurements are normalized to same precise honeycomb cell geometry.

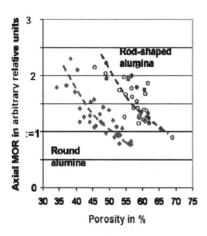
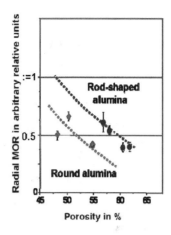

Figure 11 : Comparison of room temperature (4-point flexure) axial (left side) and radial (right side) MOR as function of material porosity of extruded honeycomb materials made from round and rod-shaped alumina particles. MOR is shown in arbitrary relative units.

DISCUSSION
Reaction Templating

Solid state reactions can occur under preservation of the precursor particle morphology. It is frequently observed for simple oxidation or reduction reaction and thin film growth, but reaction templating was also observed during more complex solid state reactions[8]. Reaction-templating is defined as a solid state reaction between a template precursor material and other reactants where the reaction product forms in or on the template with the reaction front moving from the template precursor surface into the template precursor bulk. Such a mechanism requires that the transport of the reactive matter into and through the newly formed product shell is rapid. If it is insufficient, the template particle develops a porous core and finally completely dissolves during the reaction, leaving a porous reacted shell.

Reaction-templating can be used to form a material with the precursor microstructure, but only if the precursor particle acts as a template in the solid state reaction and transforms into the final product without being dissolved or disintegrated during the solid state reaction. Templates that induce glass (or liquid phase) formation in the solid state reaction are less likely to be suited templates because of the risk of losing the original shape and forming glassy pockets. Reaction precursors that are known to transform by epitaxial or topotaxial reaction into a final reaction product are good candidates for solid state reaction-templating.

Aluminum titanate formation from titania and alumina is associated with a volume increase of more than 10%, yielding reaction-associated stresses that have to be relaxed to enable continuous product formation. For the present case of the aluminum titanate formation on alumina, it is known that diffusional transport in the aluminum titanate is much faster than in alumina, so that the aluminum titanate is expected to grow into filled grains and not into a hollow shell.

In all examples of our study, reaction templated growth of aluminum titanate was observed with a clear correlation between the surface plane of a single crystal substrate or alumina grain substrate and the grown aluminum titanate crystals. The negative expansion c-axis of aluminum titanate was always found to be preferentially oriented within the substrate plane, while the a-axis of aluminum titanate formed the preferential growth direction of aluminum titanate over a wide range of reaction temperatures.

Since anisotropic alumina particles in the powder mixtures align upon extrusion with plate faces and rod long axis preferentially pointing into the extrusion direction, the reaction-templated growth of aluminum titanate produces an overall texture in the material that directly results from the template particle alignment upon extrusion and the relative sizes of the template surfaces (aspect ratio). Figure 12 sketches examples of platy, rod-shaped and spherical grains and illustrates the interplay of template particle alignment, template particle anisotropy and aluminum titanate growth mode. Upon extrusion, rods and platelets are sheared into a preferential alignment, see Figure 12, where the extrusion direction is vertical in the image plane, For anisotropic template particles, larger surfaces contribute more to the overall aluminum titanate formation than smaller ones. Therefore, aluminum titanate is mainly formed on the rod cylinder or the large facets of the plates, so that the overall texture of aluminum titanate in the final material exhibits a preference for the a-axis to lie in the plane perpendicular to the extrusion direction and the c-axis to point into the extrusion direction.

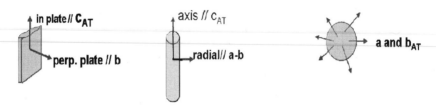

Figure 12: Schematics of aluminum titanate reaction-templated growth on alumina particles with different shape

In conclusion, forming processes such as extrusion, tape casting, cold- or hot-pressing can be used to align template particles. The alignment will be better the larger the aspect ratio. Growth on the preferentially aligned templates with a preferential growth direction or growth planes produces a preferred crystallographic orientation of the reaction product. As a result, a material is obtained that shows enhanced texture when using templates with large aspect ratios. In the studied case of aluminum titanate – feldspar composites, such texture was induced by the shape of the alumina particle shape and observed in form of a preferential alignment of the negative expansion c-axis in the extrusion direction in case of rod-shaped or platy alumina.

Thermo-Mechanical Properties Of Microcracked Materials During Temperature Cycling

Differences in thermal expansion coefficient between different phases or between different crystallographic directions within a phase can lead to microcracking during temperature changes. The microcracking is associated with a decrease in the material's elastic modulus due to a reduced connectivity of the material, changes in its thermal expansion, as well as different properties on heating and cooling (hysteresis).

In order to illustrate this behavior we will refer to a simple model/thought experiment, shown in Figure 13. The model considers two isotropic materials with identical properties except that one material has negative thermal expansion and the other has positive thermal expansion. For the sake of discussion, the positive thermal expansion coefficient is assumed to be larger than the absolute value of the negative thermal expansion coefficient. We imagine bonding equal volumes of "grains" of these two materials to two rigid parallel plates as shown in the drawings in Figure 13 at high temperature. We assume that strain relaxation occurs only by microcrack formation and that no strain-relief mechanisms, such as dislocation climb or glide, diffusion or viscous flow, are activated in the considered temperature range. The grains, cubes with side length d, are separated by empty space ("pores"). As this assembly is cooled from the initial high temperature, the two rigid plates come closer together as determined by the average thermal expansion of the two materials, and tensile stresses build in the positive expansion grain, while compressive stresses build in the negative expansion grains. The strain energy contained in the assembly is proportional to the difference in thermal expansion coefficients, ΔCTE, the amount of undercoling, ΔT, the elastic modulus, E, and the volume of material, $2d^3$. At some critical temperature during cooling, the surface energy generated by the formation of a crack in the material with positive expansion will be equal to the strain energy stored due to the difference in thermal expansion. Below this critical temperature it is energetically more favorable to form two separate grains with a separating microcrack to relieve the strain energy. This is the origin of microcracking. During the formation of the microcrack, the plates quickly separate to relieve the compressive strain in the negative expansion material. Upon further cooling, after the formation of the microcrack, further micorcrack opening is controlled by the expansion of the negative expansion material. During subsequent heating for this simplified model the plates come closer together due to the contraction of the negative expansion material until the original stress-free temperature is reached. At this point, the crack surfaces in the negative expansion material meet and can rebond. If heating is continued, the plates will then again begin to separate due to the average thermal expansion of the two materials. Similarly to thermal expansion, the elastic modulus and the thermal conductivity will go through similar changes during the heating and cooling sequence. The thermal expansion and elastic modulus of this simple model are shown in Figure 14.

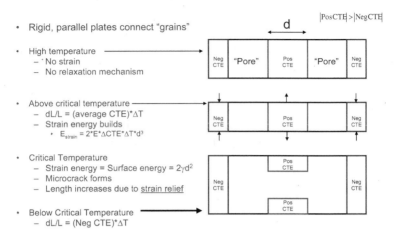

Figure 13: Simplified model system to illustrate microcracking behavior.

Discontinuities in the thermal expansion and elastic modulus behavior with temperature are determined in our simplified model by the equality of the strain energy (E_{strain}) and surface energy (γ) such that the amount of undercooling, ΔT, required for microcrack formation to be favorable is:

$$\Delta T = \frac{\gamma}{\Delta CTE \cdot E \cdot d} \cdot \qquad \text{Equation 3}$$

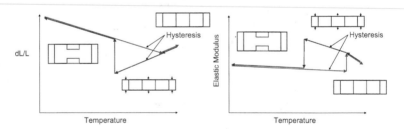

Figure 14: Temperature dependent thermal expansion and elastic modulus for a 2-grain rigid-plate model system.

Larger grains will microcrack sooner during cooling, smaller grains will microcrack later or not at all (the smallest grain that will microcrack at a given ΔT is usually referred to as the critical grain size for microcracking). Compared to our simple model, real materials will have a grain size distribution that will smooth out the abrupt changes in length and elastic modulus for the two-grain model illustrated here. The impact of this can be approximated by stacking many of these two-grain model systems with different grain sizes upon each other into a one-dimensional stack. The properties along the stacking direction will be similar to what is seen in a real material. This is illustrated in Figure 15. With a continuous grain size distribution the curves become smooth.

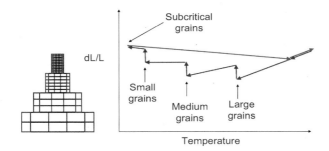

Figure 15: A stack of 2-grain model systems with a range of grain sizes illustrating the approach to a continuous thermal expansion curve along the stack direction in a system with a grain size distribution.

CONCLUSIONS

We have shown how a combination of alignment of alumina particles in simple oxide batches by extrusion and reaction-templating can be used to make aluminum titanate-feldspar composites with template-featured microstructures, anisotropic porosity and crystallographic texture. Thermomechanical properties of these materials were measured and linked to the templated microstructures. It was shown that reaction templating from alumina precursors with tailored shapes can be used to create stable, engineered microcracked materials with high porosity and improved strength compared to materials made from round alumina.

ACKNOWLEDGEMENTS

This work made use of the in-situ SEM and high temperature XRD facilities at Alfred University through the Center for Advanced Ceramic Testing (CACT). The authors want to thank R.Davis, C.Warren, R.Payrsek, M.Carson, K.Work and J.Westbrook at Corning Incorporated, S.Misture and W.Votava of Alfred University for their technical contributions.

REFERENCES

[1]G.A.Merkel, W.A.Cutler, C.J.Warren, "Thermal durability of wall-flow ceramic diesel particulate filters", *Proceedings SAE World congress* SAE 2001-01-0190 (2001)
[2]S.B.Ogunwumi, P.D.Patrick, T.Chapman, C.W.Warren, I.Melscoet-Chauvel, D.L.Tennent, "Aluminum titanate compositions for diesel particulate filters", *Proceedings SAE World congress* SAE 2005-01-0583 (2005)
[3]T.Boger, D.Rose, W.A.Cutler, A.Heibel, D.L.Tennent, "Evaluation of new diesel particulate filters based on stabilized aluminum titanate", *Motortechnische Zeitschrift* **66**(9), 2 (2005)
[4]B.Freudenberg, A.Mocellin, "Aluminum titanate formation by solid state reaction of alumina and titania single crystals", *J. Mat. Sci.* **25**, 3701-3708 (1990)
[5]H.W.Hennicke, W.Lingenberg, "Formation of aluminum titanate", *J. de Physique C1* **47**, 533-536 (1986)
[6]B.Freudenberg, A.Mocellin, "Aluminum titanate formation by solid state reaction of fine alumina and titania powders", *J.Am.Ceram Soc* **70**, 33-38 (1987)

[7]B.Freudenberg, A.Mocellin, "Aluminum titanate formation by solid state reaction of coarse alumina and titania powders", *J.Am.Ceram Soc* **70**, 22-28 (1988)

[8]T.Tani,"Texture engineering of electronic ceramics by the reactive-templated grain growth method", *J. Cer.Soc.Jap.* **114**, 363-370 (2006)

[9] B.Wheaton et al, ECERS in press

[10]J.J.Cleveland, R.C.Bradt, "Grain size/microcracking relations for pseudobrookite oxides", *J.Am.Ceram Soc.* **61**, 478-481 (1978)

[11] J.B. Walsh, "Effect of cracks on the compressibility of rock", *J. Geophys. Res.* **70**, 382-389 (1965)

[12]J.B. Wachtman, W.E. Tefft, D.G. Lam, and C.S. Apstein, "Exponential temperature dependence of Young's Modulus for several oxides," *Phys. Rev.* **122**, 1754 (1961)

WEIBULL ANALYSIS OF 4-POINT FLEXURE STRENGTHS IN HONEYCOMB CERAMIC STRUCTURES (CORDIERITE AND SILICON CARBIDE)

Randall J. Stafford, Cummins Inc., Columbus, IN, USA
Stephen T. Gonczy, Gateway Materials Technology, Mount Prospect, IL, USA

ABSTRACT

The ASTM C28 Advanced Ceramics committee has published a new standard test method (C1674) to evaluate the flexural strength of honeycomb porous ceramic materials. An interlaboratory study (ILS) was done in 2008 with over 600 specimens tested in 4-point room temperature flexure at nine laboratories. The ILS round robin study was used to develop repeatability and reproducibility data across a range of major experimental variables -- three different ceramic honeycomb structures [cordierite 1 (C558), cordierite 2 (Celcor) and silicon carbide (MSC111)] and two different flexure specimen sizes. Weibull analysis of the wall fracture strength data for these two specimen sizes (four point - ¼ point flex with 45mm and 65 mm inner spans) confirms that the larger specimen produces data with a narrower distribution [Cordierite 1 – m (large) = 16.16, m (small) – 13.02; Silicon carbide – m (large) = 16.67, m (small) – 14.92]. The differences in the Weibull strengths for the two specimen sizes in the two materials were very small (4% for cordierite 1 and 1% for silicon carbide), but with the larger specimens giving a slightly lower strength compared to the small specimens [Cordierite – σ_o (large) = 5.33, σ_o (small) – 5.52; Silicon carbide – σ_o (large) = 19.82, σ_o (small) – 20.09]. The cordierite 2 with the different micro and macrostructure gave higher Weibull statistics (Cordierite 2 – m = 14.89, σ_o – 7.64) in the small specimen size than cordierite 1 in the small specimen size. Weibull analysis of these flexure data reinforces and confirms the baseline statistical analysis (mean, and coefficient of variation) of the wall fracture strength data.

INTRODUCTION

Honeycomb substrate materials have been used in aftertreatment devices since 1975 when they were introduced for automotive catalyst supports[1,2,3]. The use of honeycomb substrates expanded to diesel applications[4] with the continuing tightening of worldwide emission regulations for nitrous oxides and particulate matter. As the applications of honeycomb ceramics have expanded, so has the expectation for performance and durability. The strength has been a key factor in providing critical information on durability and has been measured by manufacturers[5,6,7,8] and used in development of empirical models to predict survival.[9,10] More demanding applications and the advancement of technology have resulted in finite element models which require high quality data.

In 2007 the ASTM C28 Advanced Ceramics committee published a new standard test method (C1674)[11] to evaluate the room temperature flexural strength of honeycomb porous ceramic materials. In 2008 an interlaboratory study (ILS) was completed where the test standard was exercised to assess the repeatability and reproducibility of the method for two types of cordierite and one silicon carbide honeycomb ceramics. The interlaboratory study conducted over 600 flexure strength tests on the five specimen sets across multiple extrusions of each material. The initial results of that ILS were reported and published in 2009.[12]

The key conclusions of the cordierite portion of the study were 1. the equations for calculation of strength (Nominal Beam Strength, Wall Fracture Strength, and Honeycomb Structure Strength) are accurate and produce reasonable values which are comparable between laboratories; 2. the effect of lineal cell count for nominal beam strength values reported previously[6,7,8] is possible but not statistically confirmed for the tested specimen geometries and cell architectures; and 3. humidity control in the test environment is critical to producing comparable data.

This large number of tests provided an opportunity to do Weibull analysis on variations in material, specimen size, and lot-to-lot. This paper reports the results of that Weibull analysis.

EXPERIMENTAL

a. Materials

Ten (10) cordierite particulate filter substrates (commercial designation C558*) and twenty (20) cordierite catalyst substrates (commercial designation Celcor**) were provided for flexure specimen preparation. The C558 filters were 228 mm (9 inch) diameter by 178 mm (7 inch) long with a nominal 46.5 cell/cm^2 (300 cell/in^2) cell density and a cell wall thickness of 0.305 mm (0.012 inch). The Celcor catalysts were 144 mm (5.66 inch) diameter by 152 mm (6 inch) long with a nominal 93 cell/cm^2 (600 cell/in^2) cell density and a cell wall thickness of 0.102 mm (0.004 inch).

Three hundred twenty four (324) silicon carbide particulate filter substrate segments (commercial designation MSC111**) were provided for specimen preparation. The MSC111 filter segments were 35 mm (1.38 inch) by 35 mm (1.38 inch) cross section by 203 mm (8 inch) long with a 46.5 cell/cm^2 (300 cell/in^2) cell density and a cell wall thickness of 0.305 mm (0.012 inch).

b. Specimen Preparation

Five groups of flexure test specimens were prepared for flexure testing. The materials, test method, specimen cross section, linear cell count, and nominal open frontal area are shown in Table I.

Table I. Honeycomb Specimen Set Description

Honeycomb Material Sets	Manufacturer/ Designation	C1674 Method	Specimen Cross Section	Linear Cell Count	% Open Frontal Area
Set 1A Cordierite 1	NGK, C558	Method B 90 mm O-span	12 x 25 mm	9 x 17 cells	~60%
Set 1B Cordierite 1	NGK, C558	Method A 130 mm O-span	22 x 25 mm	15 x 17 cells	~60%
Set 2A Cordierite 2	Corning, Celcor	Method B 90 mm O-span	12 x 25 mm	12 x 24 cells	~81%
Set 3A Silicon Carbide	NGK, MSC111	Method B 90 mm O-span	12 x 25 mm	9 x 17 cells	~63%
Set 3B Silicon Carbide	NGK, MSC111	Method A 130 mm O-span	22 x 25 mm	15 x 17 cells	~63%

Method/Specimen A with a 65 mm inner span has 44% more outer fiber surface area under load than Method/Specimen B with a 45 mm inner span. All specimens were cut and prepared at Cummins Inc. to remove the specimen preparation as a variable in the testing program. Each of the specimens was cut to size using a fine toothed (20 tpi) hacksaw blade for cordierite material and a coarse toothed (10 tpi) steel blade bandsaw for silicon carbide material. The long faces of the specimens were hand sanded to remove most of the residual walls using 220 grit SiC paper mounted flat on a laboratory bench top.

c. Test Procedures

The flexure test specimens from each set were randomly divided into groups of twenty for distribution to the nine participating laboratories. The basic test parameters used for all specimens at each laboratory (tested in accordance with ASTM C1674, Test Method A and Test Method B) were:

- Four point ¼ point flexure geometry
- Crosshead rate of 0.01 mm/second
- Fully articulating flexure fixtures

Cordierite 1A, Cordierite 2A, and Silicon Carbide 3A specimens (12 x 25 x >117 mm) were tested at each laboratory in accordance with ASTM C1674 Method B under the additional condition:

90 mm outer span and a 45 mm inner span. (This gives a 7:1 span/depth ratio to minimize shear stresses)

Cordierite 1B and Silicon Carbide 3B specimens (22 x 25 x >150 mm) were tested at each laboratory in accordance with ASTM C1674 Method A under the additional condition: 130 mm outer span and a 65 mm inner span. (This gives a 6:1 span/depth ratio.)

Major experimental variables between the nine laboratories were ambient temperature and humidity, fixture alignment method, dimensional measurement methods, and roller diameters. Load and support bearing diameters ranged between 10 mm and 25 mm, based on laboratory availability.

The specimen mass, dimensions, and breaking force were measured and recorded for each test. The nominal beam strength and wall fracture strength were calculated for each specimen using the equations in ASTM C1674. The wall thickness and cell pitch (cell opening + one wall) were measured optically at Cummins to confirm the nominal architecture for the two honeycomb geometries.

d. Round Robin Organization

The round robin test program was conducted with nine laboratories representing honeycomb manufacturers, intermediate suppliers, product end users, academic institutions and national laboratories*** at sites in Europe, Japan and USA. The tests were conducted between October and December 2008. Not all the laboratories were able to test all five sets of specimens.

Cordierite 1A specimens were distributed to nine (9) laboratories. 158 specimens were tested with 155 specimen breaks inside the inner load span and 3 specimen breaks outside the load span. Breaks outside the load span were censored and not used in the data analysis.

Cordierite 1B specimens were distributed to eight (8) laboratories. 136 specimens were tested with 127 specimen breaks inside the inner load span, 7 specimen breaks outside the load span, and 2 specimens with incorrect cell counts. Specimens with breaks outside the load span and incorrect cell counts were censored and not used in the data analysis. (Specimens censored for incorrect cell count had either one more or one less cell in the thickness or width than the rest of the specimens tested. While this will not affect the wall fracture strength, there is a difference in the nominal beam strength, therefore to limit the number of variables under consideration, these data points were removed.)

Cordierite 2A specimens were distributed to nine (9) laboratories. 161 specimens were tested with 147 specimen breaks inside the load span, 12 specimen breaks outside the load span and 2 specimens with incorrect cell counts. Specimens with breaks outside the load span and incorrect cell counts were censored and not used in the data analysis.

Silicon Carbide 3A specimens were distributed to eight (8) laboratories. 150 specimens were tested with 135 specimen breaks inside the load span and 15 specimens with crushing of cells at the loading or support bearings. Twelve of the specimens with crushing were from one laboratory so the entire 18 specimen data set from this laboratory was excluded from the analysis. The three additional specimens with crushing were also excluded from the analysis

Silicon Carbide 3B specimens were distributed to eight laboratories. 135 specimens were tested with 83 specimen breaks inside the load span and 52 specimens with crushing of the cells at the loading or support bearings. Fifty-four specimens from 4 laboratories (45 crushed and 9 valid test breaks) were excluded from the analysis. Three laboratories (with small diameter support rollers) had no valid tests, because of crushing at the load points. An additional 7 specimens from the remaining specimen sets that showed crushing were also excluded from the analysis. One laboratory with small diameter support rollers retested new specimens using larger diameter rollers to produce valid results.

All test specimens were remeasured for dimensions to ensure that residual wall stubs on the specimen surface were not included in the dimensional measurements.

e. Strength Calculation and Statistical Analysis

The nominal beam strength, the wall fracture strength, and the honeycomb structure strength were calculated per the ASTM C1674 test method for all 629 valid test specimen results from the five test sets. From that data the means, coefficients of variation, and 95% repeatability and 95% reproducibility values were calculated for each test set.

f. Weibull Analysis Methodology

Weibull analysis of the wall fracture strength (WFS) data was done using a two parameter Weibull analysis, calculating the Weibull modulus (m) and the Weibull characteristic strength (σ_o). The analysis assumed that flexure failure was surface flaw dependent. The analysis considered strength variations related to material differences between silicon carbide and cordierite, specimen size, lot variation in each test set, and test variability between labs.

RESULTS AND ANALYSIS

a. Flexure Strength Data – Basic Statistics

Per ASTM C1674, the nominal beam strengths (NBS), wall fracture strengths (WFS), and honeycomb structure strengths (HSS) were calculated from the breaking forces, the dimensional measurements of the specimens, and the nominal open frontal area (OFA) for each of the five data sets.

Table II – Flexure Strength Averages and Statistics for Honeycomb Ceramics

Material Set	Cord 1A	Cord 1B	Cord 2A	SiC 3A	SiC 3B
Beam Size	Small	Large	Small	Small	Large
Valid Test Count	154	127	145	129	74
Lab Count	9	8	9	8	5
Nominal Beam Strength (MPa) -- Global Mean	2.05	1.93	2.60	8.30	8.10
Global Coef. of Variation	7.2 %	6.1 %	11.0 %	6.5 %	5.8 %
Coef. of Variation for the Lab Means	4.4 %	4.0 %	7.3 %	3.0 %	2.4 %
Wall Fracture Strength (MPa) -- Global Mean	5.33	5.17	7.38	19.38	19.24
Global Coef. of Variation	8.0 %	6.7 %	8.7 %	7.4 %	6.9 %
Coef. of Variation for the Lab Means	4.0 %	3.7 %	6.2 %	3.2 %	3.9 %
95% Repeatability (within Lab) (±%)	21 %	16 %	19 %	19 %	17 %
95% Reproducibility (between Labs) (±%)	23 %	19 %	25 %	20 %	20 %
Honeycomb Structure Strength (MPa) Mean	1.97	1.91	1.49	7.17	7.12

Figure 1 gives the mean values (with ± 1 standard deviation error bars) for each of the five data sets.

Major conclusions from the baseline statistical analysis are:

1. Silicon carbide (SiC) is stronger than the cordierites 1 and 2, as expected.
2. Calculating the wall fracture strength (WFS) gives a true material strength, but doesn't markedly improve the standard deviation, compared to the NBS. For specimens with these cell architectures, the wall fracture strength is ≅ 2.5X the nominal beam strength.

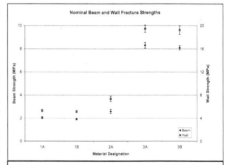

Fig. 1 - Nominal Beam Strengths and Wall Fracture Strengths for Cordierite and SiC

3. The specimen size effect for these two specimen sizes is slight and not very significant, certainly less than 1 SD on both NBS and WFS for both cordierite and for SiC.
4. The larger specimen geometry does improve the 95% confidence levels for the cordierite 1. In part this size effect reflects higher accuracy and less variability in the measurement of the dimensions of the larger specimen.
5. Cordierite 2A does have a higher wall strength than cordierite 1. It is not obvious if this is a composition or microstructure effect. But Cordierite 2A has a lower HSS, because it is an 80% porous channel structure, compared to a 60% porous channel structure for cordierite 1.
6. The 95% repeatability and reproducibility values are broad (16%-25%). Going to a larger specimen size does give some improvement (reduction) in the 95% repeatability and reproducibility values.

b. Weibull Analysis of the Honeycomb Flexure Data
    Weibull analysis of the wall fracture strength data was done for cordierite 1, cordierite 2, and silicon carbide, considering the two flexure specimen sizes, lot/extrusions variations, and variations between test labs. The Weibull (MLE) statistical results are given in Table 3.

TABLE 3 - Weibull Statistics for Wall Fracture Strength Data (Specimen Size and Lot Effects)

	Cordierite 1			Cordierite 2			Silicon Carbide		
	m	σ_o	Tests	m	σ_o	Tests	m	σ_o	Tests
Specimen Size – Small	13.02	5.52	154	14.89	7.64	145	14.92	20.09	129
Specimen Size – Large	16.16	5.33	127	Not done			16.67	19.82	74
Extrusion/ Lot A1 (Small)	14.88	5.42	56	18 extrusions with			Test specimens cut		
Extrusion/ Lot A2 (Small)	12.85	5.64	51	too few specimens			from individual		
Extrusion/ Lot A3 (Small)	13.10	5.50	47	(<10) from each			segments. No lot		
Extrusion/ Lot B2 (Large)	16.78	5.27	30	extrusion for valid			information.		
Extrusion/ Lot B3 (Large)	19.10	5.53	27	analysis					
Extrusion/ Lot B4 (Large)	17.89	5.23	32						
Extrusion/ Lot B5 (Large)	14.77	5.28	29						
Extrusion/ Lot B6 (Large)	28.22	5.37	10						

Figure 2 is a Weibull plot of the wall fracture strengths of Cordierite 1, divided into the small 90 mm outer span test data (Set 1A) and the large 130 mm outer span test data (Set 1B). This Weibull plot and the data in Table 3 show similar values for the Weibull characteristic strength (5.52 and 5.33) for the large and small test specimens. But statistical analysis shows that these two values are statistically different within the 95% confidence level.

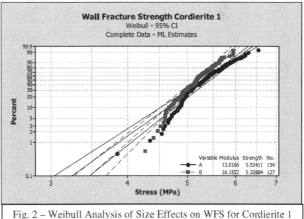

Fig. 2 – Weibull Analysis of Size Effects on WFS for Cordierite 1

The Weibull modulus values (13.0 and 16.1) are also statistically distinct at the 95% confidence level. The data for the larger specimen set (Set 1B) has a larger Weibull modulus (a narrower distribution) than the small specimens. This correlates with the coefficient of variation values in the baseline statistical analysis.

Figure 3 graphs the Weibull distribution of the strengths in Cordierite 1A (Small Specimens), showing variations in the data for specimens cut from three extrusions/lots (A1, A2, A3). All three lots have similar characteristic strengths, but there is a range for the Weibull modulus (12.8 to 14.8) which is not significant at the 95% confidence level, but does raise questions on whether lot A2 shows a bimodal flaw distribution and a difference compared to Lots A1 and A3.

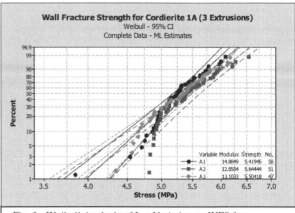

Fig. 3 - Weibull Analysis of Lot Variation on WFS for

A similar analysis was done on the extrusion/lot data from the large cordierite test bars (Cordierite 1B) in Figure 4. These larger specimens were cut from five different extrusions/lots (B2, B3, B4, B5, and B6). The five data sets again show a narrow range of characteristic strengths (5.25 to 5.53), but a broader range of Weibull moduli values (14.78 to 28.23). The high value of 28.23 can be discounted, because it is only based on 10 specimens. But the other four values have a broad enough range, that there is a question on whether there are lot-to-lot variations in the flaw distribution. Again, one lot (B3) appears to be bimodal.

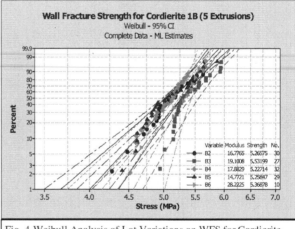

Fig. 4-Weibull Analysis of Lot Variations on WFS for Cordierite

The WFS strength data was also analyzed for differences between the nine laboratories for Cordierite 1 in the two specimens sizes. These data are plotted in Figure 5 and 6.

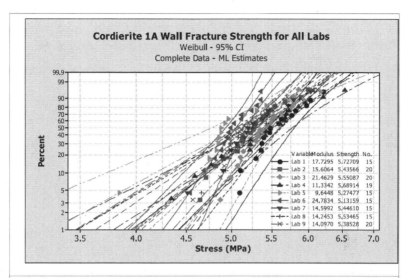

Fig. 5 - Weibull Analysis of Laboratory Variations on WFS for Small Cordierite 1A

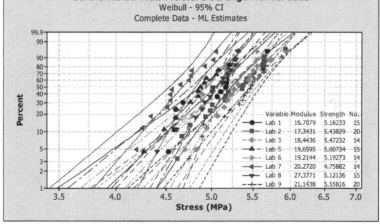

Fig. 6 - Weibull Analysis of Laboratory Variations on WFS for Large Cordierite 1B

There is one significant difference between the data for the small and large specimens. The small specimens have lower Weibull modulus values (9.6-24.7), compared to the large bars (16.7-27.3). This indicates a wider distribution of strengths for the small bars.

These graphs show the variation between laboratories, both in terms of the Weibull modulus and the characteristic strength. Certain laboratories (#5 and #6) show low characteristic strength values and/or low Weibull modulus values. These variations will be discussed in the section on humidity effects.

Figure 7 shows the Weibull analysis of the WFS data for the Cordierite 2A small specimens. Since there is only one size specimen in this set, there are no size effects to evaluate. Cordierite 2A can not be directly compared to Cordierites 1A and 1B as it has different macro- and microscopic structures.

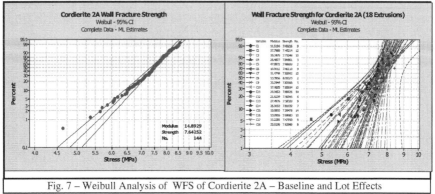

Fig. 7 – Weibull Analysis of WFS of Cordierite 2A – Baseline and Lot Effects

The second graph in Figure 7 shows the Cordierite 2A data divided into the 18 different extrusion/lot sets. There is very wide variation in the Weibull modulus, the characteristic strength, and the 95% confidence bounds here, primarily because the data sets are small (less than 12).

The WFS strength data for Cordierite 2A were also analyzed for differences between the nine laboratories. These data are plotted in Figure 8, showing the Weibull moduli and the characteristic strengths for the different laboratories. Certain laboratories again show low characteristic strength values and/or low Weibull modulus values. These variations will be discussed in the section on humidity effects.

c. Humidity Effects on Cordierite

All the laboratories reported the lab environment testing conditions – temperature and relative humidity. The temperature range was narrow -- 21°-26°, but the relative humidity range was large – 15-75%. The wall fracture strength values from all the cordierite specimens were analyzed for the interaction effects of the temperature and humidity. Statistical analysis, by ANOVA, of each material sample set

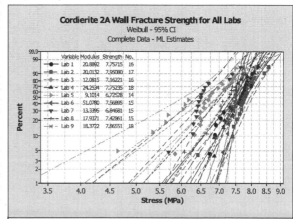

Fig. 8 - Weibull Analysis of Laboratory Variations on WFS for Small Cordierite 2A

showed that temperature was not a significant factor in the strength differences between laboratories.

The humidity effects for Cordierite 1A, 1B, and 2A are shown in Figure 9 where the mean WFS values of each lab set are plotted against ambient humidity with ± 1 standard deviation error bars. The laboratories with the higher humidity had lower strength data for the cordierite.

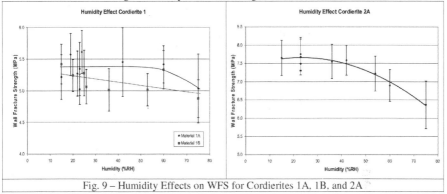

Fig. 9 – Humidity Effects on WFS for Cordierites 1A, 1B, and 2A

The ANOVA analysis did show that humidity was a significant interaction factor for Cordierite 1A (weak interaction) and Cordierite 2A (strong interaction). The humidity effect was not as strong for the larger 1B specimens.

These data indicate a possible humidity effect that points to hydrolysis corrosion and slow crack growth, reducing the strength of the cordierite, particularly for Cordierite 2A. This has been observed in other work on cordierite[13]. The humidity effect may occur not only during testing, but also in uncontrolled storage with a possible exposure time effects. Humidity exposure may be a critical variable for cordierite mechanical properties and deserves further investigation.

d. Weibull Analysis of the Silicon Carbide Honeycomb Flexure Data

Figure 10 is a Weibull plot of the wall fracture strength data for the silicon carbide, divided into the small 90 mm outer span test data (Set 3A) and the large 130 mm outer span test data (Set 3B).

This Weibull plot and the data in Table 3 show similar values for the Weibull characteristic strength (20.09 and 19.82) for the large and small test specimens. Statistical analysis shows that these two values are distinct within the 95% confidence level. The Weibull modulus values (14.92 and 16.67) are also statistically distinct at the 95% confidence level. The larger cross section specimen data set (Set 3B) has a larger Weibull modulus and a narrower distribution, than the small specimens. This correlates with

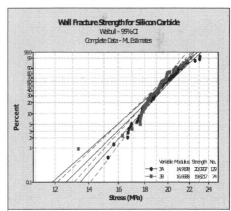

Fig. 10 - Weibull Analysis of Size Effects on WFS for Silicon Carbide (3A and 3B)

the coefficient of variability values calculated in the baseline statistical analysis.

Since each section of silicon carbide honeycomb produced one or two test bars, there was no opportunity to evaluate lot-to-lot variability.

The WFS strength data was also analyzed for differences between the nine laboratories for silicon carbide in the two specimens sizes. These data are plotted in Figure 11.

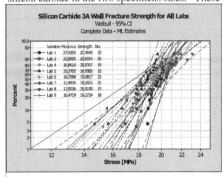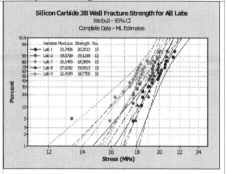

Fig. 11 -- Weibull Analysis of Laboratory Variations on WFS for Silicon Carbide 3A and 3B

There is one significant difference between the data for the small and large specimens. The small specimens have a wider range of Weibull modulus values (11.5-27.4), compared to the large bars (15.1-27.7). This indicates a wider distribution of strengths for the small bars, as also seen in the global analysis.

These Weibull graphs show the variation between laboratories, both in terms of the modulus and the characteristic strength. The characteristic strength values with a range of 18.4 to 20.5 are relatively uniform between labs and between the two specimen sizes. However, there is a broader range of Weibull modulus values between labs and between sizes, with m values for 11.5 to 27.4 for the small specimen and m values for 15.1 to 27.7 for the large specimens. There is no obvious explanation for the difference in standard deviations between the different laboratory sets.

An analysis of the silicon carbide WFS mean values against humidity values for the different laboratories was inconclusive, as shown in Figure 12. If there is a humidity effect, it is relatively minor (<10% loss) under these testing conditions.

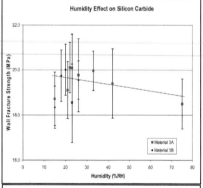

Fig. 12 – Humidity Effects on WFS for Silicon Carbide

CONCLUSIONS

1. Weibull analysis of the wall fracture strength data for these two specimen sizes (four point - ¼ point flex with 45 and 65 mm inner spans) confirms that the larger specimen produces data with higher Weibull modulus and a narrower distribution [Cordierite – m (large) = 16.16, m (small) = 13.02; Silicon carbide – m (large) = 16.67, m (small) = 14.92].

2. The differences in the Weibull strengths for each material in the two specimen sizes were very small (4% for cordierite 1 and 1% for silicon carbide), but with the larger specimens giving the

lower strength [Cordierite – σ_o (large) = 5.33, σ_o (small) = 5.52; Silicon carbide – σ_o (large) = 19.82, σ_o (small) =20.09].

3. The cordierite 2 with the different micro and macrostructure gave higher Weibull statistics (Cordierite 2 – m = 14.89, σ_o = 7.64) in the small specimen size than cordierite 1 in the small specimen size.

4. There are possible lot-to-lot variations in cordierite 1 but these differences are relatively minor.

5. There is variation in the Weibull parameters between laboratories for cordierite, but it appears that much of this variation is a possible humidity effect. Humidity exposure may be a critical variable for cordierite mechanical properties and deserves further investigation.

6. Weibull analysis of the silicon carbide data shows lab-to-lab differences in Weibull but there is no obvious explanation for these differences. Humidity is not an obvious cause of these differences.

Weibull analysis of these flexure data reinforces and confirms the baseline statistical analysis of the data.

FOOTNOTES

* NGK Automotive Ceramics USA Inc, Novi, MI
** Corning Inc, Corning, NY
*** Caterpillar Inc, Mossville, IL; Corning Inc, Corning, NY; Cummins Inc., Columbus, IN; Dow Chemical Co., Midland, MI; Johnson Matthey, Wayne, PA; NGK Automotive Ceramics USA Inc, Novi, MI; Oak Ridge National Laboratory, Oak Ridge, TN; Robert Bosch Gmbh, Stuttgart, Germany; Università di Padova, Padova, Italy

ACKNOWLEDGEMENTS

We appreciate the support of NGK Ceramics and Corning Inc and providing the test material that was instrumental in conducting this round robin program. Also, the diligence and persistence of Leigh Rogoski and Ken Rogoski at Cummins Inc. in preparing, packaging and shipping the specimens to the different laboratories for testing.

REFERENCES
[1] P.M.Then, P. Day, The Catalytic Converter Ceramic Substrate – An Astonishing and Enduring Invention, *Interceram (Germany)*, Vol. 49, No. 1, 20-23 (2000).
[2] R.D. Bagley, R.C. Doman, D.A. Duke, R. W. McNally, Multicellular Ceramics as Catalyst Supports for Controlling Automotive Emissions, *SAE Paper 730274* (1973).
[3] J.S. Howitt, Thin Wall Ceramics as Monolithic Catalyst Supports, *SAE Paper 80082* (1980).
[4] J. Adler, Ceramic Diesel Particulate Filters, *International Journal of Applied Ceramic Technology*, Vol. 2, Issue 6, 429-439 (2005).
[5] S.T. Gulati, R.D. Sweet, Strength and Deformation Behavior of Cordierite Substrates from 70° to 2550°F, *SAE Paper 900268* (1990).
[6] S.T. Gulati, K.P. Reddy, Size Effects on the Strength of Ceramic Catalyst Supports, *SAE Paper 922333* (1992).
[7] J.E. Webb, S.Widjaja, J.D. Helfinstine, Strength Size Effects in Cellular Ceramic Structures, *Ceramic Engineering and Science Proceedings – Mechanical Properties and Performance of Engineering Ceramics II*, Vol. 27, Issue 2, 521-531 (2006).
[8] R.J. Stafford, R. Wang, Effect of Test Span on Flexural Strength Testing of Cordierite Honeycomb Ceramic, *Proceedings of the 31st International Conference on Advanced Ceramics and Composites,* J. Salem and D. Zhu, Ed., The American Ceramic Society, Westerville, OH, (2007).

[9] J.D. Helfinstine, S.T. Gulati, High Temperature Fatigue in Ceramic Honeycomb Catalyst Supports, *SAE Paper 852100* (1985).

[10] S.T. Gulati,Long-Term Durability of Ceramic Honeycombs for Automotive Emissions Control, *SAE Paper 850130* (1985).

[11] ASTM C1674, Standard Test Method for Flexural Strength of Advanced Ceramics with Engineered Porosity (Honeycomb Cellular Channels) at Ambient Temperature, *Annual Book of ASTM Standards,* Vol 15.01, ASTM International, West Conshohocken, PA, (2008).

[12] R.J. Stafford , S.T. Gonczy, Data Reliability for Honeycomb Porous Material Flexural Testing, *Proc. 33d Int. Conf. on Advanced Ceramics and Composites, Symp. 9 – Advanced in Porous Ceramics,* Edit. D. Singh, J. Salem, Am. Ceramic Society, (2009).

[13] K.Kaneko, Triple-layered, Thick Glassy Grain Boundaries in Porous Cordierite Ceramics, *Acta Materialia,* Vol. 50, Issue 3, Pages 597-604 (2002).

[14] ASTM E691, Standard Practice for Conducting an Interlaboratory Study to Determine the Precision of a Test Method, *Annual Book of ASTM Standards,* Vol 14.02, ASTM International, West Conshohocken, PA, (2009).

Author Index

Author Index